3650

INTRODUCTION TO FISH TECHNOLOGY

Joe M. Regenstein
Professor of Food Science
Department of Food Science,
Cornell University, Ithaca, New York

Carrie E. Regenstein
Assistant Director for Instructional and Personal Systems,
Cornell Information Technologies

An Osprey Book
Published by Van Nostrand Reinhold
New York

An Osprey Book
(Osprey is an imprint of Van Nostrand Reinhold)

Printed in the United States of America.

Van Nostrand Reinhold
115 Fifth Avenue
New York, New York 10003

Chapman and Hall
2–6 Boundary Row
London SE1 8HN, England

Thomas Nelson Australia
102 Dodds Street
South Melbourne 3205
Victoria, Australia

Nelson Canada
1120 Birchmount Road
Scarborough, Ontario MIK 5G4, Canada

Cover by Beth Kochen

16 15 14 13 12 11 10 9 8 7 6 5 4 3 2 1

Library of Congress Cataloging-in-Publication Data
Regenstein, Joe M.
    Introduction to fish technology / Joe M. Regenstein
and Carrie E. Regenstein.
        p.   cm.
    "An Osprey book."
    Includes bibliographical references and index.
    ISBN 0-442-00500-8
    1. Fishery technology.   I. Regenstein, Carrie E.   II. Title.
SH334.7.R44   1991
    338.3'727—dc20   90-22556
                                                        CIP

For our parents

Contents

Preface

"Fish" can be finfish, shellfish (molluscan or crustacean), or any other form of marine or freshwater animal life that can be used for human or domestic animal consumption. The emphasis of this book is on the teleost—or bony—fish that are used as food for humans.

Fish obtained either from the wild or from farms (aquaculture) requires proper handling from the moment of catch until consumption. Training in how best to accomplish this task must be provided to both those doing the day-to-day work and those responsible for improving and modernizing the system through new developments in technology, science, and engineering. This book is offered as an introduction to fishery food science for both future fish technologists and scientists and those in the industry who want to learn more about the exciting industry in which they are participating. The authors believe that students must go beyond the narrow limits of their field to understand the many cross-currents and interactions that will affect the outcome of their efforts. Sociology, politics, economics, psychology, and history affect what can be successfully accomplished by the applied food scientist or technologist working in the fishing industry.

Twenty percent of the world's fish can be found off the eastern end of North America, and an even larger percentage off the west coast. In the coming years the United States will try better to exploit its fishery resource. Although the descriptions and explanations in these chapters emphasize the North American fishery, the authors hope that much of what is said will also be helpful to others around the world.

Because of our desire to draw on a wide range of materials, we have attempted to integrate many facets and levels of fishery information. The early chapters should give the reader the background information and vocabulary that will be needed for the ideas that will be developed in more detail in the later chapters.

It is hoped that this book will be read by people in the fishing industry interested in improving the quality of seafood as well as by students in food science or related fields. For many readers, we hope it may serve to illustrate how chemistry, biology, microbiology, and engineering might be integrated and applied to the study of a single commodity area. We hope that current fish industry workers may find the book to be a handy summary and reference resource of much otherwise scattered information.

1

Background

Contrary to popular belief, most of the sea is relatively barren. Most harvestable marine resources are close to the shore within the region encompassed by the continental shelf, that is, in relatively shallow water. The ability of any animal to grow in the water column depends on the availability of appropriate conditions for such growth, particularly proper nutrients, appropriate temperature and salinity conditions, and the absence of toxic substances. Aquatic plants, especially phytoplankton at the bottom of the food chain, depend on photosynthesis for obtaining energy. All other life depends directly or indirectly on this energy obtained from the fixation of carbon dioxide as carbohydrate, except for some special bacterial processes that are not currently of importance to the commercial fishing industry. These photosynthetic organisms are generally limited to surface depths that can still receive some sunlight (light energy). Other nutrients are often obtained from the run-off of rivers and streams. Thus the coastal estuaries, which receive these run-offs—and are brackish water rather than full-salinity ocean water—often serve as the breeding grounds for many commercially important seafood species.

An important natural oceanic phenomenon for providing nutrients for large fish populations is the upwelling phenomenon. The various sea currents churn up the ocean bottom and increase the nutrient content of the water column. In a fairly constant-temperature environment, often maintained for long distances, these currents become extremely productive fishing areas. But if these currents change, the available stocks can disappear rapidly. The "El Niño" currents of the Pacific Ocean off North and South America have been quite erratic, and the temperature and location changes have had a very negative impact on fisheries from Peru to Alaska. El Niño reflects the appearance closer to shore of an increased water temperature, which is detrimental to the traditional species living within the normally colder inshore waters.

Another example of fickle currents that play an important role in the recruitment of young fish occurs in the Northwest Atlantic, that is, just off

1

the United States' northeast coast. The shelf water current is closest to shore and generally moves from north to south; as such, it is subject to wide temperature fluctuations. The slope water current is further out. This current runs from north to south along the continental shelf and tends to maintain a fairly warm temperature. Furthest out is the Gulf Stream, which moves from south to north. Part of this stream sometimes breaks off into the slope current, forming warm "core rings" which can last for a long time before slowly moving back down the slope current. These break-out currents can be up to 320 km (200 miles) in diameter and move clockwise. Those formed on the other side of the current in the deep ocean move counter-clockwise. If these warm core rings appear over the fish spawning and larvae growing grounds at the wrong time, they move the larvae out beyond their normal grounds and the affected fish cannot be recruited into the fishery.

Within the continental shelf region, how well are we currently harvesting the sea's limited resources? By weight, the Food and Agricultural Organization of the United Nations (FAO) indicates that the current yearly harvest of fish from the wild is about 100 million tons (1 metric ton = 2,200 lb), of which almost 15 million metric tons are from aquaculture; between 75 and 80% is used directly for human food; the rest are industrial fish, that is, fish that are used for the production of fish oil, fish meal, and fish solubles mainly as an ingredient in animal feeds (Chapter 10). Approximately 88% of the world catch is finfish, 8% is molluscs, and 4% is crustaceans.

Estimates suggest that between 60 and 70% of the maximum sustainable yield (MSY, that amount of fish that can be harvested without harming the resource) of the oceans' 120 million metric tons is currently harvested. Thus, at best, we cannot even double the yearly production from the oceans without a negative impact on total fish resources. Some fishery biologists believe that we are already harvesting a larger percentage of the total available resource: FAO has lowered its estimate of the total annual potential to 100 million metric tons; and the United States National Marine Fisheries Service (NMFS) quotes that figure as 80 million metric tons. If more than the maximum sustainable yield for a particular fish is harvested, there will be fewer fish available in subsequent years. Unfortunately, many important fisheries such as cod, pollock, haddock, and flounder seem to be in this predicament.

The total wild fish harvest in 1948 was about 20 million metric tons. Of the total current harvest, about 9 million metric tons are from aquaculture and predictions for the realistic future go as high as 30 million metric tons. Thus, aquaculture may provide a significant portion of the world's fish supply in the coming years.

Less than 5% of the total worldwide protein supply is obtained from the sea so that, at best, the potential from natural fishery stocks is less than 10% of the world's protein need. In the past few years the catch worldwide has remained relatively stable. Furthermore, the FAO in 1981 estimated that 28 million metric tons of fish were caught from stocks that were fully exploited or actually depleted. In recent years the problem has got worse.

This does not mean that the sea's proteins do not play an extremely important role in meeting the food needs of many people, particularly for those people fortunate enough to live close to the sea. In many developed and underdeveloped countries fish could play a much larger role if the available fish stocks were fully and properly exploited at their maximum sustainable yield and if the stocks once harvested could be more fully used, that is, if the waste and spoilage could be eliminated or at least significantly reduced. A fish stock that is not harvested is essentially lost. In many cases, fisheries biologists managing stocks try to deal with entire fish populations such as the Atlantic groundfish, so that the various interspecies interactions might be taken into account to yield the best overall harvest level.

Other questions arise concerning stock assessment. How many fish are out there? What is the maximum sustainable yield for a single species? What is the optimum yield for a total biomass? The numbers needed for these decisions are difficult to obtain and the resulting values are often disputed. No working fisherman believes the figures obtained by the biologist who takes a few cruises each year and fishes in a few places, usually far from the working fisherman. Yet, intelligent management of our fishery resources depends on the accuracy of these numbers. In some fisheries, fishermen are beginning to participate in the data collection process, improving both the quality of the data and the fishing community's acceptance of the results.

There are also social and political concerns related to fishery resources. Who stands to benefit? Who stands to lose with each management decision? For example, since 1976 Canada has stated that the welfare of the people of the Maritime provinces was as important a concern in formulating fishery policy as considerations of the biological stock.

FISHERY MANAGEMENT IN THE UNITED STATES

The United States' management of the marine fisheries has been assigned by law to the regional fishery management councils. These bodies are supposed to represent local and state fishery interests as well as national interest. Though made at the federal level, appointments are based on suggestions from the individual states. The final management plans must be approved by and carried out by the National Marine Fisheries Service (NMFS) of the National Oceanographic and Atmospheric Administration (NOAA) of the Department of Commerce (DOC). The management councils and the NMFS often differ in their attitudes toward the fisheries and implementations of management activities. The interaction between these two groups has caused friction between them and may at times hinder fisheries management. But it does give a greater say to those directly affected by the decisions being made.

Expanding resource utilization in underdeveloped countries requires cultural and psychological sensitivity: Consider an individual fisherman in a Third World country, for example. If you give him or her a bigger boat, a stronger motor, new equipment; will he or she catch more fish? Can the infrastructure for handling and distributing fish absorb these additional fish? Interestingly, the answer to the first question is often "No." Some fishermen quit fishing when they catch a certain number of fish or when they have earned enough money to meet their immediate needs. Once that is accomplished, they will come back to shore to "enjoy" life.

POLLUTION

Development of the coastal areas for other human uses, such as recreation and housing, usually renders these areas ecologically unfit to perform their role in developing fisheries resource. Unfortunately, such development also brings other problems: most notably, pollution. One example is the large canyons that are part of the continental shelf. These canyons are part of the habitat for lobster, red crab, and tilefish. But, in the New York Bight, a designated region of the Atlantic Ocean, off the coast of New York and New Jersey, there is concern over the dumping of sewage sludge at sea. The federal government now requires that all sewage sludge-dumping take place at a site over 100 miles from the coast instead of the previously designated area 12 miles offshore. Animals caught near the new dump site are already showing signs of various skin diseases. The Corps of Engineers (a unit of the United States Army), however, is still permitted to dump dredging spoils, that is, the materials removed when a channel or harbor is enlarged, at the closer site. (These spoils often contain various organic and mineral contaminants.)

There are two methods commonly used by the government to monitor the presence of sludge in the marine environment. One method is to determine the presence of *Clostridium perfringens*; this bacterium is used because the spores survive the salt water. The other is to measure the quantity of tomato seeds in the water—a sure sign of human waste!

MARINE MAMMALS

Another issue that has become significant in recent years is the question of marine mammals such as seals and dolphins. These animals are generally protected by various laws. However, they eat lots of fish and are also the alternate host for the seal worm (see Chapter 14) that may be found in the flesh of fish such as cod. For many years, the Canadian government permitted a managed hunt for baby gray seals in the Maritime provinces. This humane slaughter of seals by a quick blow to the head allowed the size of the herd to be kept at a reasonable level. These seals are *not* on any endangered species list. However, organizations like Greenpeace and the International Fund for Animal Welfare have used this hunt to generate sufficient publicity that the European Economic Community (EEC) has banned the import of seal pelts—essentially closing the seal fishery because seal was no longer economic to hunt. After a few years, however, a Canadian Royal Commission has decided that a cull of the gray seal population is essential, as they are estimated to cause $60 million worth of damage to other fish. New alternatives are being sought for seal control such as birth control and deworming. Apparently, the bulk of the worms come from the old bulls, who often congregate on Sable Island, Nova Scotia, where they could be more easily treated after being tranquilized. At this time the Canadian Ministry of Fisheries has banned all killing of baby seals.

For an excellent study of this complex issue, we strongly recommend the book by Henke (1985).

FISH BIOLOGY

Finfish are part of the phylum Chordata, subphylum Vertebrata. Within the subphylum there are three classes of fishlike animals. The first is the Cyclostomata ("round-mouthed") which includes the lampreys, hagfishes, and slime eels. These are the only living vertebrates without jaws. Chondrichthyes ("cartilaginous-fish") make up the second class. Sometimes called "Elasmobranchii," it includes the sharks, dogfishes, skates, and rays. The whale shark is the largest true fish. Its entire skeleton is cartilaginous. These fish have sharply pointed placoid scales that are firmly attached to the skin and have tiny spinous projections.

The third class of fish includes the Osteichthyes ("bony-fish"). These fish have a true bone skeleton and any one of several types of scales, specifically ganoid, ctenoid, or cycloid scales. Ganoid scales are heavier and thicker than the other types of scales; ctenoid scales have minute spiny projections at their exposed edges; and cycloid scales lack the minute spines and have rounded edges. These fish may also be assigned to the class Pisces. There are various subclasses. The subclass Palaeopterygii includes sturgeon and similar fish. Most of the true fish are in the subclass Neopterygii, which now includes what used to be the Teleoi (Teleost) class. The third subclass,

Crossopterygii, includes the coelacanths, which for many years were thought to be extinct, and the lungfishes.

A gutted teleost fish generally is about 73% flesh, 21% bone, and 6% skin. In filleting fish, the actual recovery (edible yield) is about 35 to 40% of the total weight. Some other processes have slightly higher yields, but a great deal of flesh is wasted by the process of filleting. For example, estimates of flesh yield from fish cooked and eaten whole indicate yields of 60 to 65% edible flesh.

Most fish breathe through gills. The end-product of their protein metabolism, ammonia, is excreted through the gills directly into the water. However, high levels of ammonia in the water supply block excretion by limiting the diffusion of the fish's ammonia and are therefore toxic to most fish. Ammonia levels are therefore an important consideration in establishing water quality standards for aquaculture.

A benefit of ammonia excretion to a fish is an increase in the caloric (energy) return derived when it metabolizes protein. While higher animals only recover 4 calories/gram of protein when they synthesize urea or other excretory compounds from the ammonia, fish get 5.4 calories/gram for the metabolized protein. Because fish are also cold-blooded, they do not require energy to maintain body temperature, but are very sensitive to environmental temperature. Feed efficiency with fish can reach one pound of feed per pound of gain, almost double that of the most efficient land animal—the broiler chicken. Note, however, that feed efficiency comparisons can be misleading because they fail to take into account the moisture or protein contents of the feeds, which can affect their cost.

Although they are underutilized in many parts of the world, sharks are used for food, especially in the Orient. Their dried fins contain a large amount of collagen, which is ideal for thickening and flavoring soups. Sharks have a great deal of urea in their blood, apparently for the purpose of regulating the blood's osmotic state with respect to the external seawater (osmoregulation). Urea is generally considered to be a protein denaturant. Shark's blood also contains a high level of trimethylamine oxide (TMAO). This compound is a protein structure maintainer and its high levels in the blood may have evolved to counteract the denaturing influence of the urea. Notice that both compounds are nonionizable; they do not interfere with the fish's electrolyte balance.

If the urea is not immediately removed from sharks when they are caught, the urea is rapidly broken down into ammonia. If the shark is properly bled on deck, however, the level of urea can be significantly decreased. This is generally done by cutting off their tails. In some countries, a darker (redder) color of flesh is desired with the smaller shark species like dogfish, so bleeding is not carried out. In either case, rapid chilling is required. Because dogfish are a relatively small fish, it is possible to get a rapid enough cooling of the entire animal so that ammonia production is minimized.

The wings of skates and rays can be eaten. Many people liken their flavor and texture to that of scallops. One occasionally hears unconfirmed rumors of skate wings being transformed (i.e., stamped-out with a cookie cutter!) into "scallops." Given the size of the skate harvest and the size of the scallop catch, the amount of skate "scallops" would be quite small—even if all of the skate wings harvested were converted into scallops.

Lampreys are one of the few parasitic vertebrates, and are considered by many to be a nuisance fish, in part because they are capable of destroying other fish life. Efforts to develop food usages might be appropriate, but will be hindered by their unusual activity, that is, still wriggling four to six hours after their heads are cut off!

Fish can be divided into a number of different categories depending on the fact(s) chosen for emphasis; for example, some fish are relatively lean, that is, low in fat. The normal seasonal/sexual yearly cycle of these fish shows a marked change in the protein/water relationship in the course of the year. During the spawning season, when the bulk of their biological activity is directed towards the build-up of reproductive organs (roe or milt), the flesh may be depleted of protein, that is, muscle tissue mass, and replaced by water. Thus, the quality of the fish (firmness) and its nutritional value may be minimal at this point. There is a report from the Torry Research Laboratory in Aberdeen, Scotland, of a fish harvested just post-spawning (i.e., just after releasing either its row or its milt) whose total body composition was found to be over 95% water. Subtracting skin and bones, this does not leave much room for protein. Roe—and especially the milt—are unfortunately not generally recovered for food in many countries such as the United States and may represent up to 10 to 15% of the fish's body weight.

The fatter fish show a fluctuation in their fat/water ratio as they go through their yearly biological cycle. An extreme case is menhaden ("bunker"), the major United States industrial fish, which can vary in fat content from 2 to 25% in the course of its yearly cycle.

The livers of lean and fatty fish show interesting and opposite properties, that is, the fatty fish tend to have lean livers while the low-fat fish tend to have fatty livers. The liver oil of low-fat fish such as cod is high in the fat-soluble vitamins A and D. Fish oils were once an important source of these vitamins prior to the development of commercial organic synthesis of these compounds. With the resurgence of an interest in fish oils for use in human foods, specifically for their potential therapeutic uses, interest in recovery of fish livers may again become viable. Note: some customers must be careful to avoid an overdose of vitamin A.

Fish can also be categorized by the color of their muscle. All fishes, like other animals, have two major types of muscle. The darker "slow" muscles are generally found only along the lateral line in fish. (See

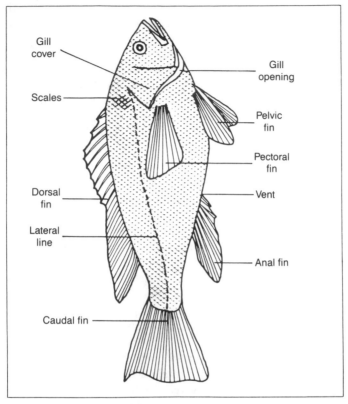

Figure 1.1 External appearance of a round fish. (*Faria, 1984, p. 181; Courtesy of Massachusetts Division of Marine Fisheries*)

Figures 1.1 through 1.6 for general schematics of fish structure; both a round fish and a flatfish are shown along with a scallop.) These are the "aerobic" muscles, that is, they have a high myoglobin content allowing them to retain oxygen, and are used for the fish's continual swimming activity. The percentage of this muscle is greater in the more migratory (pelagic) fish. The rest of the fish's muscles are "fast" or burst-type, and they are used mainly for quick escapes from danger. In some fish the white muscle contains so little myoglobin that the muscle is really white when the fish is bled. If the fish is not bled, the muscle may be pink in color due to the remaining hemoglobin from the blood. The extra hemoglobin may be beneficial nutritionally, but may also provide iron for greater bacterial growth and, therefore, more rapid spoilage. However, bleeding for fish is recommended in most Western countries because whiteness is perceived by consumers to

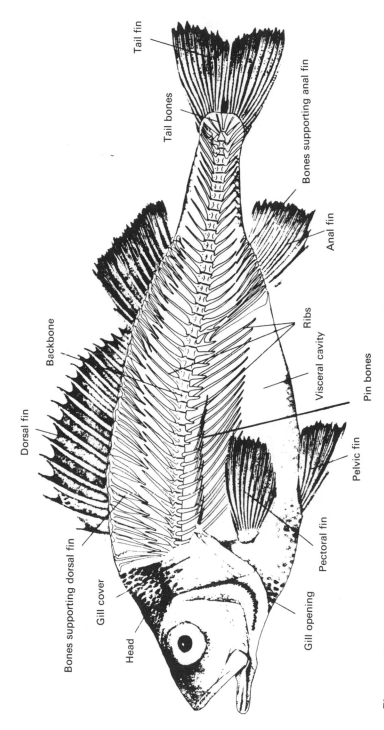

Figure 1.2 Cutaway appearance of a round fish. (*Faria, 1984, p. 124; Courtesy of Massachusetts Division of Marine Fisheries*)

Tail fin

Tail bones

Bones supporting anal fin

Anal fin

Backbone

Dorsal fin

Ribs

Visceral cavity

Pin bones

Pelvic fin

Bones supporting dorsal fin

Gill cover

Head

Pectoral fin

Gill opening

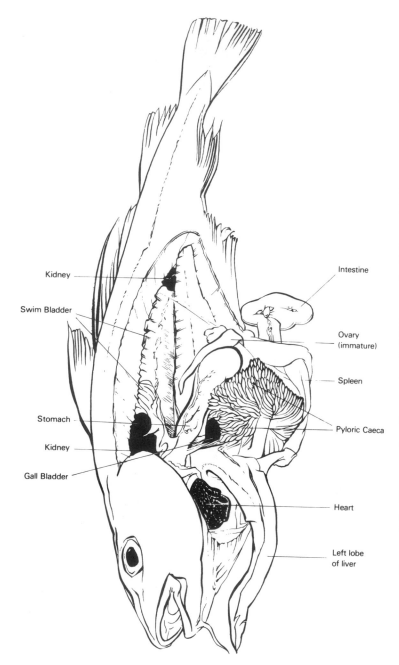

Kidney

Swim Bladder

Stomach

Kidney

Gall Bladder

Intestine

Ovary
(immature)

Spleen

Pyloric Caeca

Heart

Left lobe
of liver

Figure 1.3 Diagram of a dissected cod. (*Aitken et al., 1982, p. 11; Crown copyright*)

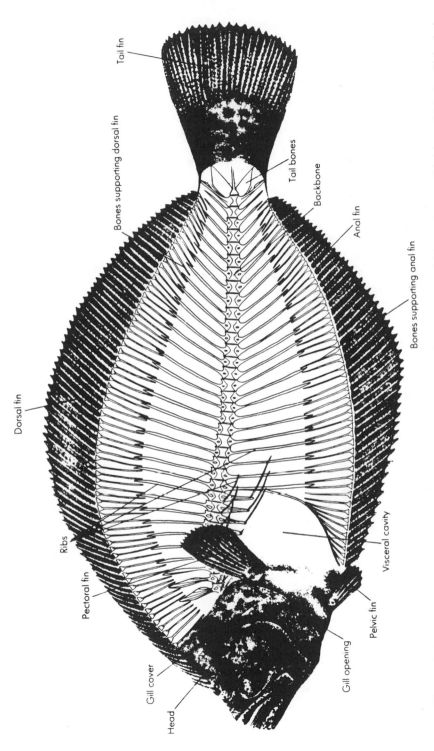

Figure 1.4 Cutaway appearance of a flatfish. (*Faria, 1984, p. 125; Courtesy of Massachusetts Division of Marine Fisheries*)

Tail fin

Bones supporting dorsal fin

Tail bones

Backbone

Anal fin

Bones supporting anal fin

Dorsal fin

Visceral cavity

Ribs

Pectoral fin

Pelvic fin

Gill opening

Gill cover

Head

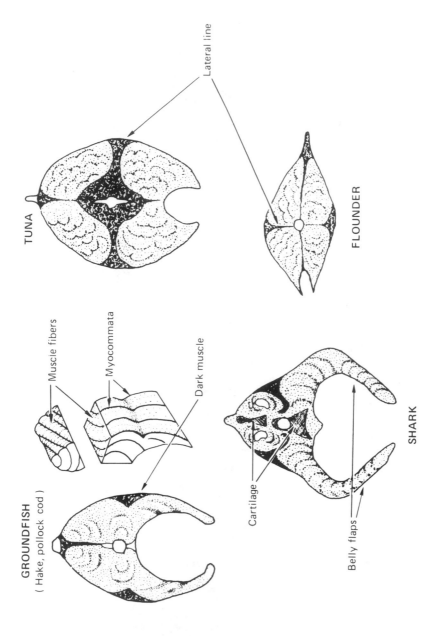

Figure 1.5 Transverse sections of various fish species. (*Martin et al., 1982, p. 16*)

TUNA

Lateral line

FLOUNDER

GROUNDFISH
(Hake, pollock cod)

Muscle fibers

Myocommata

Dark muscle

SHARK

Cartilage

Belly flaps

12

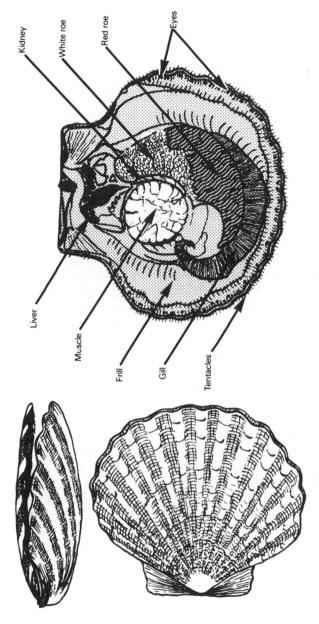

Figure 1.6 Diagram of a scallop. *(Aitken et al., 1982, p. 129: Crown copyright)*

Kidney

White roe

Red roe

Eyes

Liver

Muscle

Frill

Gill

Tentacles

be desirable in fish. Those fish that do a lot of swimming may also have more myoglobin in these muscles and are therefore darker in comparison to their nonswimming brethren.

The color of fish flesh may also be due to the presence of other pigments. These are generally obtained from the food eaten by the fish and are lipid-soluble carotenoid pigments that are absorbed and deposited in the flesh. Shellfish, unlike most finfish, can synthesize these compounds and are often the source of these colors in finfish that consume shellfish. Duplicating the colors of wild fish in aquacultured fish has often proven difficult because of the absence of enough of such pigments in the diet. Recent work on the extraction of pigment from shellfish waste (especially crayfish) may eventually lead to a solution of this problem.

Location of feeding area creates another category with which to describe fish. Demersal fish are bottom feeders and pelagic fish are generally surface and/or plankton feeders.

Another distinction can be made between freshwater and saltwater fish, whose body chemistry and, thus, post-harvest handling differ. A few fish are anadromous, that is, during their life cycle, they switch from fresh to salt water, and back. Eels and salmon are in this category, the latter being the most important commercially. Some fish grow mainly in the brackish water found in bays and estuaries, which is not as salty as regular seawater.

Some categorization of nonfinfish species might also be useful. Shellfish are generally either crustaceans, for example, lobster, shrimp, crab, and crayfish; or molluscs, for example, oysters, scallops, clams (bivalves), squid, cockles, and welks. Echinoderms, for example, sea urchins and sea cucumbers (Beche Le Mer) may also be used for food, particularly in the Orient. Marine mammals such as whales, seals, and walruses are harvested in many parts of the world; in certain countries, however, they are protected species. Their proliferation has had a negative impact on both the quality and quantity of shellfish and finfish available for human harvest. A recent University of Rhode Island study estimates that marine mammals—not including seals—consume between 365,000 and 545,000 metric tons of food in the Northeast region of Canada and the United States. Of this amount, 185,000 to 276,000 metric tons were fish, 154,000 to 224,000 metric tons were squid, and 26,000 to 45,000 metric tons were zooplankton. These animals eat large quantities of commercially valuable seafood. Moreover, they are often, as mentioned earlier, the alternate host for marine parasites that can be found in finfish, for example, the seal worm.

WHO CATCHES FISH?

Who has the right to catch fish? This is often a controversial question. Fish are generally considered to be a common property resource, that is, they are available to anyone willing to go after them. In practice, many governments require some sort of license before allowing fishing. These licenses require a fee, but the fees are generally not high enough to discourage entry into the fishery. Neither is there a restriction on the number of licenses issued. In the United States most states require a license to fish in freshwater (for those over a certain age), but there is currently no requirement for a license for saltwater sports fishing on the East coast. The West coast, however, has introduced saltwater licenses. Commercial fishermen, on the other hand, generally do have to obtain a license.

Licenses for sports fishermen give them more visibility. The fees also give the state more incentive to develop programs favorable to sports fishing.

In some countries and/or specific fisheries, "limited entry" to the fishery resource is a major issue. If the resource is limited, does the government have a right to take away a "common property" and allow only certain people to benefit from the resource? In practice, almost all fishery management schemes do this indirectly, that is, by "categories" rather than by limiting individual fishermen, but this does not constitute a limited entry program. On what basis should the share allocation of the resource in a limited entry program be fixed: by the boat or by the person? How are the rights to be transferred to others? These controversial questions have often hindered the successful development of these programs.

Most managed fisheries set quota maxima for fish species in terms of particular types of gear, size of boat, fishing grounds, allowed fishing time, and so on; but it is then left to the competitors within the fisheries to find and catch the fish. This often leads to a concentration of effort at the beginning of the quota period. In some fisheries this has led to four- to six-hour openings, for example, to catch halibut! The overabundance of boats and people usually means that no one can make a decent living. Solving this problem often falls to the social welfare system.

On the other hand, the use of a limited entry system leads to a better spreading of the catching period and a more rational capitalization of the fishing fleet. Boats can operate more efficiently and more attention to quality is possible. Inevitably, a few fishermen will in fact "strike it rich." Limited entry in the United States fishing industry can be found in certain fisheries in a number of states. Other agricultural marketing systems, for example, the Canadian egg marketing system, contain similar provisions for supply management. However, many fishermen consider such systems undemocratic, a violation of their basic right to fish, and many people feel that the system almost always favors those who are already in the business to the detriment of potential newcomers.

As a rule, most of the major fishing grounds used to be legally on the "high seas," that is, not under any country's jurisdiction. However, abuses of this system, for example, overfishing by some of the countries with large freezer-trawler fleets, led to significant changes. Freezer-trawlers are large boats, usually over 200 feet long, that both fish and further process the fish on the same boat. They may also be referred to as "factory" ships.

During the 1970s, many countries extended their coastal fishing territory ("extended jurisdiction") from 12 to 200 miles off their coast to bring most of the important fish stocks within their legal control and increase the economic return from their adjacent continental shelf. Almost all coastal countries now claim a 200-mile jurisdiction, often using their small un-inhabited islands to greatly extend the territory being claimed. In recent years, some countries have been discussing moving their jurisdiction out to 300 miles. For example, a small but important part of Canada's fishing area is beyond the 200-mile limit. Canada feels that foreign countries are fishing this region to the detriment of its own citizens.

Naturally, disputes have occurred where the territories claimed by different countries overlap. One of the most controversial of these was the United States/Canadian fisheries border in the Atlantic Ocean. The case went to the World Court, which essentially split the disputed territory. Both countries are unhappy with the decision. What effect the free trade agreement recently signed by Canada and the United States will have on fisheries remains to be seen (see Chapter 16).

Fishing on the "high seas" without legal constraints has changed in other ways, too. Highly migratory fish that roam the vast expanses of relatively barren sea are now either regulated actively (i.e., by some type of international agreement) or the fishing is limited in scope because of political/social/economic pressure. For example, Atlantic Ocean tuna are now regulated; there is also an international whaling treaty. Catching salmon in the Pacific (high seas interception using gill nets) is now regulated, although Japan and other key countries do not subscribe to the relevant treaty. Recent economic and political pressure on Japan, the major purchaser of salmon, and other nonmembers has limited the violations of this treaty by nonsubscribing countries. However, the National Marine Fisheries Service has become more energetic in enforcing the laws dealing with importing illegally caught salmon into the United States.

With extended jurisdiction, it has become necessary for foreign fishermen to obtain fishing rights for a particular fishing area before commencing fishing operations. For countries that had depended heavily on their distant-water fishing fleets, this has led to major economic and/or food supply problems. Countries such as Japan, Poland, West Germany, Spain, Portugal, Russia, and Korea have had to accept many constraints on their large freezer-trawler-based fleets that used to roam and fish the seas freely.

For those limited foreign fisheries permitted by the United States government, an American observer is required on board the vessel at the foreign fisherman's expense. This arrangement is obviously fraught with physical, political, and personal hazards to the observer. In some fisheries, particularly in Alaska, the state is requiring observers on domestic boats.

Access to the relatively rich United States fishing grounds by other countries is also a very political process (see Chapter 16).

ALTERNATE USER GROUPS

In many of the more developed nations the availability of ocean frontage has become limited and the price for waterfront land has gone up significantly. The alternate uses, for example, tourism and housing-related activities, are often better able to afford higher land costs. These groups may then not consider the presence of a commercial fishing fleet with old and smelly boats as a "romantic" neighbor.

The most common alternate users along the coasts are the sports fishermen, many of whom believe that commercial fishermen catch too many of the available fish. Sports fishermen are often able to gain a political advantage over commercial fishermen because of their greater numbers in a "one-man, one-vote" society. On the other hand, commercial fishermen, although fewer in number, are trying to feed all of society's citizens. A further discussion of this conflict appears in Chapter 16.

Despite the conflict between sports and commercial fishermen, a common interest in having fish available can sometimes lead to a unified effort to improve a fishing site. For example, there is currently a joint effort to protect the ecology of Chesapeake Bay now that it appears that acid rain is a serious problem rather than just overfishing. Unfortunately, short-term conflicts between these two user groups sometimes prevent effective mobilization against long-term dangers of encroachment such as at-sea pollution by, for example, oil and mineral operations.

For some species of fish, the recreational catch is not trivial. For example, in New York State the commercial catch of bluefish is about 7 million pounds, while the recreational catch is 99 million pounds. The bluefish run in later summer and early fall, a nice time for sports fishermen. New York State law has for many years set the commercial bluefish catch quota at about 15% of the recreational catch. It can usually be documented that the average sports fisherman spends more money per fish than the commercial fishermen. Commercial fishermen are quite concerned that many species of fish may not be available to them in the future, for example, redfish and striped bass.

The large amount of fish caught by recreational fishermen leads to other anomalies. This catch is not generally included in the various food consumption figures. Thus, evaluations of the diets and health status of various populations may be distorted by a failure to consider this sometimes major source of food/protein. On the other hand, recreational fishing may take place in more marginal waters, particularly polluted freshwater streams and lakes. Thus, some states have had to issue advisories concerning the eating of fish from various waters; this has taken the form of outright bans or the establishment of weekly limits, especially for pregnant women, children, or those who might be immunocompromised. Some states have recognized that some of the lower-income populations will continue to eat these fish anyway; education programs have been started to teach people how to handle fish to minimize the amount of contamination ingested.

New York has initiated several programs in this area of education. For example, migrant community leaders (e.g., in the fruit growing areas along Lake Ontario) teach proper fish handling to their followers in their native dialect. The focus has been on how to remove the fish's skin, dark meat, and belly flaps (the regions containing the largest amount of fat); and how to cook the fish, for example, baking on a rack to remove even more fat. The majority of the contaminants of concern are fat-soluble. Smaller fish also seem to have lower levels of contamination than larger fish. This, of course, conflicts with the management needs to set size minima on many fish species and the general "consumer desire" for bigger fish.

The publicity associated with the contamination and pollution of areas fished by sports fishermen has spilled over into the commercial supply. Many consumers do not realize that the commercial supply is much more carefully monitored and in general has fewer problems with contaminants.

Another concern with respect to the recreational catch is its role in undocumented economic activity (i.e., the selling of fish, usually for cash, by "sports" fishermen). This also discounts their role as retail competitors with the commercial fishery. There is a difference between catching fish for one's own use and selling a dozen bluefish every day of the season to a local restaurant for cash. Other problems can follow; for example, it is the commercial catch that is more closely monitored for contamination, not the sports catch.

REFERENCES

Aitken, A., I. M. Mackie, J. H. Merritt, and M. L. Windsor. 1982. *Fish Handling and Processing*. Edinburgh, Scotland: Her Majesty's Stationery Office.

Faria, Susan, M. 1984. *The Northeast Seafood Book*. Boston: Massachusetts Division of Marine Fisheries.

Henke, Janice Scott. 1985. *Seal Wars: An American Viewpoint*. St. John's, Newfoundland: Breakwater Books Ltd.

Martin, Roy E., George J. Flick, Chieko E. Hebard, and Donn R. Ward. 1982. *Chemistry and Biochemistry of Marine Food Products*. Westport, CT: Avi.

2

How Fish Are Caught

The method used to catch fish affects the condition in which the product is landed. In most developed countries the process of catching fish is successful; the limited availability of fish worldwide is usually due to problems in handling, processing, or distribution of fish. Marketing can also be difficult, especially for underutilized fish; these used to be called "trash" fish, but that name seems inappropriate for fish that food technologists believe people ought to be eating. Let us explore a few of the many methods used to catch fish.

NET METHODS

Trawling

For many of us, the first image that comes to mind when we think of "commercial fishing" is a boat dragging a large net through the water. This technique is called trawling. *Midwater trawling* is when the net is somewhere in the water column; it is called *bottom trawling* when the net is dragged along the ocean floor. Many of the most important fish species are caught with the bottom trawl. Large "doors" are used ahead of the net (Figure 2.1) to keep the net down yet opened wide.

There are many designs of these doors. The most common are the "otter doors," the largest of which may weigh 3,400 kg (1,545 lb). Recent efforts have focused on reducing a trawl door's drag to cut down on fuel consumption. One trick has been to "cut holes" into the doors in appropriate places. Trawling requires a relatively strong boat to pull all this equipment through the water—and fish, too! It uses a lot of fuel; it tends to disturb the bottom of the ocean; it tends to trap undersized fish; and a number of fish are lost, or "dropped," if the bottom, or "cod-end" of the net is accidentally opened to release the fish. However, it is still amazingly efficient at catching large quantities of bottom fish.

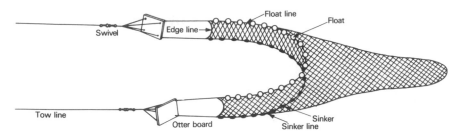

Figure 2.1   An otter trawl net. (*Yamaha Motors, 1986, p. 134*)

The dust cloud sent up by the trawl doors appears to be an important factor in how these nets catch bottom fish. It is not clear at this time what effect the absence of the dust cloud from the trawl door can have on the success of bottom pair trawling (see next paragraph).

The doors used for mid-water trawls are much lighter than those used for bottom trawling. But in either case, if one used two boats, each dragging one side of the net, one could accomplish the same opening of the net without using any doors. Doing this, of course, takes cooperation between two boats (and boat captains are notoriously independent people). There are also some technical constraints on the horsepower and handling characteristics of the two boats. Because pair trawling is so comparatively fuel efficient, it can sometimes increase the profitability of both boats' fishing effort without the necessity of doubling the catch of the two individual boats.

But what is the condition of fish that is trawled? The fish from one trawl (up to 36 metric tons (80,000 lb) on the larger offshore boats, for example) are all landed on the deck of the boat at once. This creates a great deal of physical pressure on the fish during the landing. In addition, rocks and other abrasive material (for example, starfish) may be picked up from the ocean floor and included in the net haul. The fish are crushed and rubbed against each other; worse, they may also be subjected to abrasion by materials that are much harder than other fish.

On deck, the crew are faced with a large amount of fish. The natural desire of the crew, who are paid by the value (price per pound times the weight of the catch), is to process the fish as fast as possible. Even with their best effort, the last fish processed from a tow may be much older than the first. An extreme situation occurs with the British midwater mackerel fishery: when fishing is good, it might take almost 24 hours to process the fish caught in a half-hour tow. This can be disastrous in the summertime for mackerel, a relatively rapidly spoiling fish. (Chapter 14 includes a discussion of histamine (scromboid) poisoning.)

Although trawling clearly cannot yield the highest quality of fish, it is still used extensively because of its efficiency. Shorter tows with smaller amounts of fish are now recommended. An unfortunate reality is that quality is not always the most important consideration in the fishing industry, although the attitude does seem to be changing for the better.

The ability of smaller fish to escape is supposedly controlled by the size of the net's mesh. However, as the bag fills up with fish, and the larger fish block the net holes, small fish become trapped deep inside the net and are unable to escape. Regulating the mesh size of the net may, therefore, be only partially effective in minimizing the catch of undersized fish. Because the possession of these small fish is generally illegal and they are difficult to market, they are thrown overboard. Rarely do they survive and, thus, they become a total loss, both for food or as future reproductive stock. Recent research suggests that the shape of the net mesh might affect its efficiency; a square mesh may actually allow for better escape of undersize fish than the traditional diamond mesh. When the net is being towed, only two sides of the square mesh are under tension. This causes less damage to fish escaping through the net. Normally there is a high mortality rate for fish escaping through the diamond mesh. Unfortunately, a square net is more difficult for fishermen to mend.

Table 2.1 shows some data on the relationship between fish length and weight. Table 2.2 relates the mesh size of a net to the length of fish retained. The lengths given represent the size at which about 50% of the fish are retained and 50% can still make it through the net. The data may also be

Table 2.1 Optimizing the Catch by Regulating Mesh Size. Length–Weight Data for Some Common Fish

	Weight (lb)		
	1	2	4
		Length (inches)	
Pollock		20	
Cod	14	19	25
Haddock	15	18	26
Whiting	15	20	
Redfish	12	15	20
Mackerel	12	16	
Yellowtail	15	18	
Rainbow trout	14		

Data courtesy of the National Marine Fisheries Service.

Table 2.2 Mesh Size Versus Retention Length for Some Common Fish

	Mesh Size (inches)				
	4	$5\frac{1}{4}$	$5\frac{1}{2}$	$5\frac{3}{4}$	$6\frac{1}{4}$
	Retention Length (50% retained)				
Cod	13.5	17.7	18.6	19.4	22.0
Haddock	12.9	17.0	17.8	18.6	21.0
Yellowtail Flounder	8.7	11.4	12.0	12.5	14.2
Pollock	13.2	17.3	18.2	19.0	21.4
Winter Flounder	8.4	11.0	11.5	12.0	13.6
Dabs (plaice)	9.4	12.3	12.9	13.5	15.2

Data courtesy of the National Marine Fisheries Service.

expressed for a particular fish as shown in Table 2.3. When discussing mesh size, we are referring to the mesh size of the main part of the net and not the "cod-end." As the fish are funneled to the catching or "cod-end" of the net, they enter a smaller mesh area.

The regulation of mesh size has been an important component in the overall regulation of the United States fisheries. Fisheries managers are very interested, then, in just what happens inside the net during the fishing process. Researchers at the Marine Laboratory in Aberdeen, Scotland, have taken motion pictures of trawl nets in action. These films indicate the importance of the fishes' swimming speed and the time it takes fish to become fatigued:

Table 2.3 Relative Rates of Retention for Cod

Length (inches)	Percent retained	
	$5\frac{1}{2}$ Inch Mesh	6 Inch Mesh
27	99	97
25	97	88
22	78	56
21.6	74	50
20	53	29
19.8	50	27
18	27	11
16	9	3

Data courtesy of the National Marine Fisheries Service.

as the fish tire, they fall into the cod-end of the net. Analysis of these films should eventually help in improving net designs.

The popularity of trawling has been expanded with the development of factory ships, for example, ships that can accommodate further processing and/or freezing of fish on-board. In some cases the ship also does the fishing. In other cases the factory ship may receive its fish from smaller trawlers. The smaller boats transfer their net to the mother ship with the fish still in the water. This method is generally used for joint ventures between the United States and foreign fisheries (see Chapter 16). The catching boat, obviously, needs at least two sets of nets.

For the first time, the United States is building large trawlers for ground fishing that will occur in Alaska and on the East Coast. The high cost of these boats (over $10 million each) is partly due to the Jones Act (a regulation passed in the late 18th century) which requires that all boats fishing in United States territorial waters have hulls that were made in the United States.

Until recently, most United States trawlers were less than 100 feet long. The longest in Alaska is the 295-foot Arctic Trawler. On the East coast the largest ship to date had been the Amfish at 219 feet. The Amfish had 38,000 cubic feet of freezer space and was designed to process 40 tons of fish per eight-hour day using a 40-man crew taking trips of three to four weeks' duration. However, the boat has been moved to Alaska after a short, uneconomic effort at sea in the East. Presently no large catcher-processors are operating on the United States East coast. In Canada, the two largest fishing/fish processing companies, National Sea Products (Nat Sea) of Nova Scotia and Fishery Products (FP) of Newfoundland have each been given (and are using) permission by the Canadian government to operate one large factory-trawler.

In 1985, the Alaska Factory Trawlers' Association represented four companies with a total of nine factory ships. More boats have since come on-line in Alaska both within the organization and without. These vessels had the capacity to produce 20 million pounds of whitefish fillets (mostly pollock). For comparison, the United States imports whitefish fillets from Canada (90 million pounds), Iceland (42 million pounds), Denmark (16 million pounds), and Norway (7 million pounds). "Whitefish" is a term used to describe many different types of white, low-fat, bland tasting marine fish such as cod, haddock, whiting, and pollock.

Seining

Seine nets are often used to catch schooling fish (Figure 2.2). Pelagic fish, for example, mackerel and menhaden, are most easily caught this way. These are upper-water-column fish that swim large distances. A small boat, or

Two-boat purse seining net

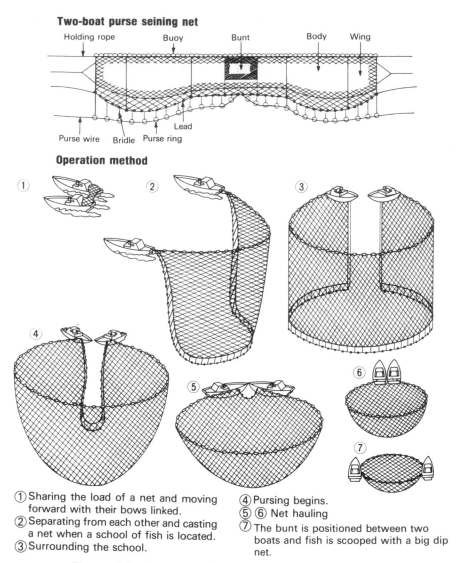

Operation method

① Sharing the load of a net and moving forward with their bows linked.
② Separating from each other and casting a net when a school of fish is located.
③ Surrounding the school.
④ Pursing begins.
⑤ ⑥ Net hauling
⑦ The bunt is positioned between two boats and fish is scooped with a big dip net.

Figure 2.2 Seine nets. (*Yamaha Motors, 1986, p. 178*)

sometimes two boats, sets off from the mother ship and carries a relatively small-mesh net with it that is set around the school of fish. If the net is a "purse" seine, a string on the bottom is then pulled tight so that the bottom of the net can be closed off. The net is then hauled in until most of the fish are in a small area of water. This is called "hardening of the net."

The fish are then generally "brailed" out, that is, removed from the water, with hand nets or with mechanical unloaders (pumps). Large quantities of fish often need to be handled rapidly, once again leading to some sacrifice in quality; the pumps themselves often damage the fish. However, the greatest damage is done to the fish at the time of unloading from the boat because of the postmortem physiological changes that have occurred in the fish. Fish in *rigor* (technically in *rigor mortis*) should generally not be pumped. Seining is used for industrial fish; the largest reported single net set catch was 250,000 lb. The boat needed help from other boats in the area to handle and stow the catch!

A special problem arose with the seining of tuna. These fish often congregate below schools of dolphins, which would also get trapped in the nets. New nets have now been designed so that the dolphins can escape or be released before the tuna are hauled in. The effort to redesign the nets, to make an extra effort to allow the dolphins to escape, and the willingness of the industry to change to these more expensive systems was a direct result of public pressure. Other countries must use these nets in order to export tuna to the United States. However, the issue remains controversial and some restaurants are not serving tuna because of concern for the killing of dolphins.*

A modification of the seining method is often practiced from shore. The "mothership" becomes some form of land-side attachment (people, vehicles, or solid object) and the net, after setting, is hauled to shore by either people or a vehicle. This method is often used for more valuable species such as bluefish, striped bass (e.g., in New York) or for salmon (e.g., Scotland).

Gillnetting

Gillnetting involves stretching out a net to intercept fish (see Figure 2.3). The net can be set at any level in the water column by manipulating the rope length and by the use of anchors and floats. Placement is often chosen to intercept natural fish migrations. The fish do not see the net and simply swim into it. Once their gills go through the net, they can no longer escape, and they drown. The thrashing of the fish prior to death leads to both external and internal bruising. Also, if the nets are not well attended, the dead fish will be untreated for too long and will be of poor quality. For this reason, gillnetting is one of the least desirable ways to catch fish from a quality standpoint, even though it is extremely efficient. It is an easy method to use with small boats; for example, the gillnet fishery for salmon in Alaska—

---

* To clarify another common misconception: mahi-mahi, or *dolphin* fish, is a regular fish and is not related to marine mammals such as dolphins.

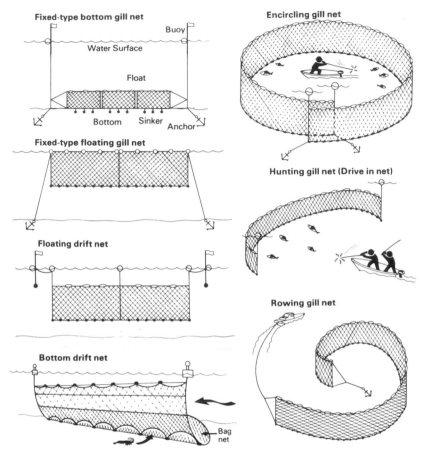

Figure 2.3 Examples of gill and drift nets. (*Yamaha Motors, 1986, p. 50*)

the major fishery during the summer spawning run—is restricted to boats under 34 feet. The size, mesh, and placement of the nets while fishing can be controlled by the regulatory agencies. The nets are generally hauled from one side of the boat and immediately sent back to fish on the other side, or a drum is used to wrap the net around as fish are simultaneously removed by hand.

In Alaska, the distance from the mouth of each river at which the nets can be placed *each day* affects the total catch efficiency of the fleet. The state uses the system to control salmon escapement. The further from the mouth of the river that nets must be placed, the greater the escapement. When sufficient salmon pass into the stream, the fishery managers permit the fishermen to harvest all they can. Spawning salmon are easier to catch closer to the river's mouth or when already in a stream. However, as they proceed up river, the quality of the salmon declines because they are not eating. Following spawning, Pacific salmon die; Atlantic salmon, however, can return to sea.

Gillnetting has become controversial because of what is called "ghost fishing." With modern net materials like nylon, these nets are difficult for fish to see, and may continue to catch fish after the fisherman has lost them. Underwater photography, however, suggests that the amount of ghost fishing being done by these nets is relatively insignificant. They tend to "ball" up and thus become more visible, while taking up less area.

Cast Nets

Weighted "cast" nets, or variations thereof, are usually used by the artisan fisherman. These are simply thrown out to sea from land or a small boat; the weights around the net cause them to sink and trap (catch) the fish.

Herding Devices

Another net-based catch method involves the use of traps, and weirs (see Figure 2.4). These devices "herd" fish into an area where they can then be

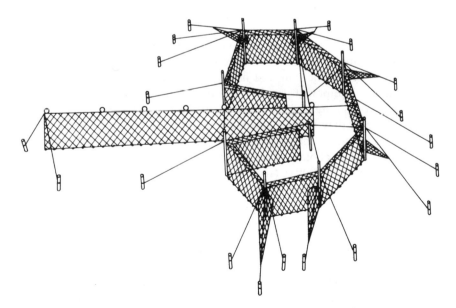

Figure 2.4 A small fixed net. (*Yamaha Motors, 1986, p. 67*)

easily removed. This fixed gear is expensive and time-consuming to build, and also depends on the fish coming to the trap rather than a boat's going to the fish. Thus, traps work best when the migration of the desired fish species can be predicted accurately. They often work best in rivers or streams, or close to shore where the natural terrain features may naturally channel fish movement. The fish can be caught live simply by hand-netting them out of the trap. Often, because they are close to shore, the travel distance/time may be minimal. Furthermore, the fish can be kept alive in the traps for some time and harvested when needed.

Large "pound" nets are used along the shore to collect and harvest these fish. The leads can be very long (e.g., almost a mile long in Chesapeake Bay) and a single net can be very costly (e.g., over $10,000).

HOOKING METHODS

Longlining

Longlining is simply the use of a long fishing line with lots of hooks, usually evenly spaced (e.g., every 3 feet) along the line (see Figure 2.5). This method can be used on rougher grounds (bottoms) which could not be used by a trawl. This system has many important benefits over trawling, including the potential for size and species selectivity by proper choice of hook size and shape, and bait; lower energy costs; and less damage to ocean bottoms.

Placing the bait onto the numerous hooks is very labor-intensive. New forms of mechanical or semi-mechanical baiters have been developed that replace the tremendous effort involved in hand baiting. It is also labor-intensive to remove the bait and fish from the line as it is hauled back on board. Again, equipment to do this automatically has been developed. However, if longlining is done properly, the fish can be processed onboard one at a time as they come aboard, often still alive.

The hooks used for longlining must be of the appropriate size for the fish that is sought. The choices of hook size and bait can be used to make this form of fishing very selective (see Figures 2.6 and 2.7). Also, since each hook is designed to catch fish above a certain size, this technique is particularly effective in avoiding undersized fish. Hook shape may also affect species selectivity as well as the amount of bait needed (a major expense with longlining). It is essential that the line(s) is stored on deck in such a way that it does not become tangled as the line is fed into the water.

The Mustad company of Norway has developed the most advanced of the numerous longline systems now appearing on the market. This system is designed to allow as many as 30,000 hooks on one line to be fished per day (28 miles of hooks!). Seabank Industries of Boston built three 76-foot

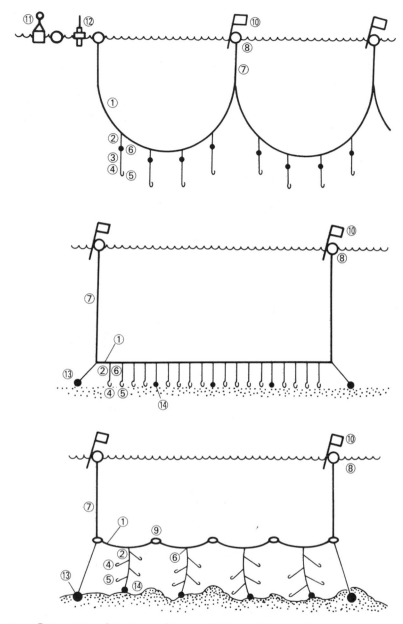

① Trunk line ② Branch line ③ Sekiyama ④ Snood ⑤ Hook ⑥ Swievel ⑦ Buoy line
⑧ Float ⑨ Anti-pressure float ⑩ Marker flag ⑪ Marker light ⑫ Radio buoy ⑬ Anchor
⑭ Sinker

Figure 2.5 Examples of longlines. Top, drift longline. Center, bottom longline. Bottom, vertical line bottom longline. (*Yamaha Motors, 1986, p. 219*)

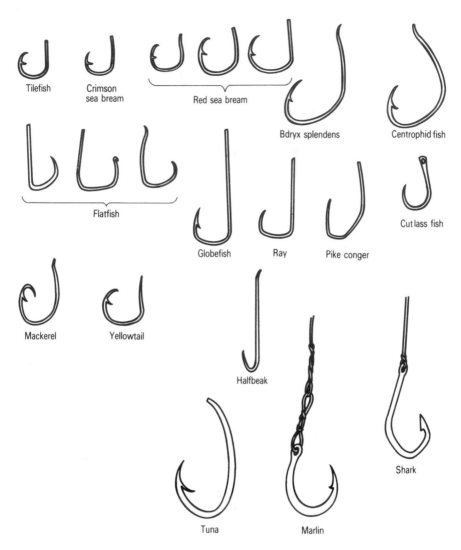

Figure 2.6   Various fishing hook shapes. (*Yamaha Motors, 1986, p. 220*)

boats—each for a crew of nine people—specifically designed to use this advanced fishing system. The boats were successful in catching fish, but the company was unable to keep up with the mortgage payments and has since gone bankrupt. An unfortunate part of that experience was the apparent attempts at sabotage of the equipment, possibly because the traditional fishing industry was opposed to the introduction of this radically new technology. The company is trying to restructure to return to fishing.

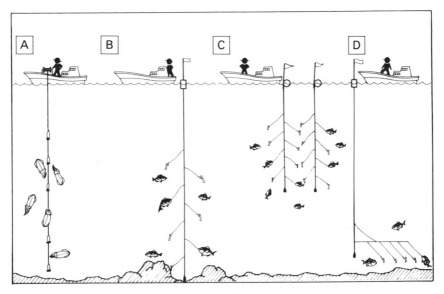

Figure 2.7   Use of vertical lines. (*Yamaha Motors, 1986, p. 211*)

Trolling

Trolling is usually done close to the water's surface. The boat moves through the water with a number of lines, each attached to a pole (see Figure 2.8). A number of hooks may be attached to each line. The fish are almost always landed live. Trolling allows for selective handling of fish and thus leads to a very high quality. Troll-caught salmon are particularly prized, especially if they are to be smoked subsequently. The absence of either internal or external bruising maximizes the appearance and the recovery (weight) of this extremely expensive product. As with all of these methods that yield higher-quality fish, it must also be assumed that the subsequent handling after catching is consistent with the higher initial quality.

Jigging

Jigging was first developed commercially in Japan. When fish are caught by this method, bait (chum) and/or special lights (at night) are used to attract the fish to an area (see Figure 2.9). In their feeding frenzy, the fish tend to bite on any flashing item including the many hooks present in the water. Some part of the catch may also be snagged on the hooks. The system of powerful lights on the boat (to create a shadow in the water) coupled with brightly colored jigs and small in-the-water lights is particularly attractive

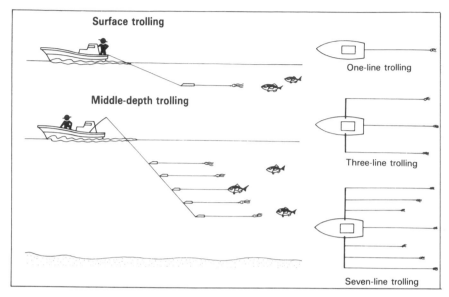

Figure 2.8　Examples of trolling. (*Yamaha Motors, 1986, p. 211*)

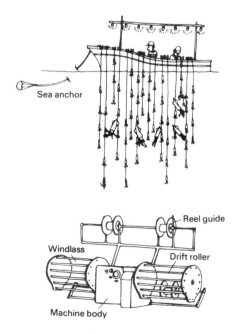

Figure 2.9　Automatic squid fishing jigs. (*Yamaha Motors, 1986, p. 24*)

to certain fish, for example, certain species of squid. These jigging lights may be battery-operated or may use the new chemical luminescence tubes, that is, chemically produced light.

Bait is an important cost in all the hooking methods, particularly longlining. Each longline hook uses about 45 g (0.1 lb) of bait. Therefore, a great deal of interest exists in developing appropriate baits, that is, those that can withstand the physical constraints of the mechanical system and also attract the appropriate fish, while using materials cheaper than bait fish (e.g., fish waste). Research is also currently being done on the "taste" preferences of fish, that is, flavor attractants. This research would be used to develop appropriate artificial baits. Another approach has been to use "feathered" hooks without bait, similar to the type of lures sports fishermen use.

OTHER TECHNIQUES

A number of more unusual techniques may also be used. Spears and harpoons (the latter, sometimes with a bomb implanted in them) may be used for some of the larger fish. In the Mediterranean, fishermen sometimes simply detonate a bomb in the water and then collect all of the dead fish—usually of low quality. Some fisheries use divers to catch live fish of higher quality.

A wide variety of other catch methods are used for shellfish. These animals often live below the ocean floor and thus require techniques like surface-dredging (e.g., of scallops) (Figures 2.10 and 2.11), or raking and tonging to recover the product (Figures 2.12 and 2.13). A dredge has steel teeth that scrape and stir-up the ocean floor. The scallops are then caught in a large steel bag behind the dredge. Lots of detritus is also picked up by the bag and a high discard rate must be expected. Tongs are really a pair of "forks" with a center pivot that are used to pick up clams or oysters from the ocean floor.

Classical rod and reel fishing can be used commercially for some of the larger (100 lb or more), more valuable fish such as tuna and swordfish.

There are many other methods used to catch fish as well as a variety of other types of equipment such as pots and traps for shellfish (see Figure 2.14). Most of these methods have been adapted over time to the specific fish and local fishing conditions. Modern research often suggests that local designs come quite close to matching anything the technologists can devise.

An important resource for testing nets and other fishing gear is a flume tank. Using this, the behavior of model nets can be studied at various water flow rates.

The fishing method and the amount of fish caught sometimes determines subsequent handling of harvested fish—which is the topic of the next chapter.

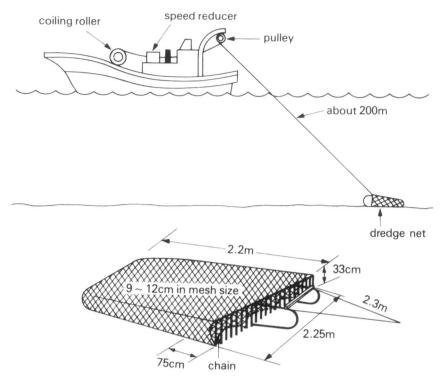

Figure 2.10 Operation of a dredge. (*Yamaha Motors, 1986, p. 75*)

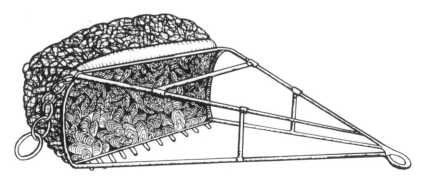

Figure 2.11 An oyster dredge. (*Wheaton and Lawson, 1985, p. 95*)

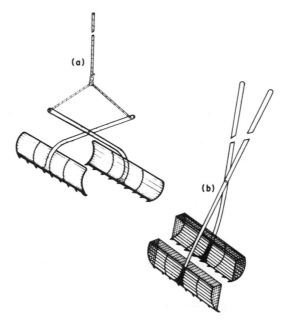

Figure 2.12 Oyster tongs. (a) Patent tongs. (b) Hand tongs. (*Wheaton and Lawson, 1985, p. 100*)

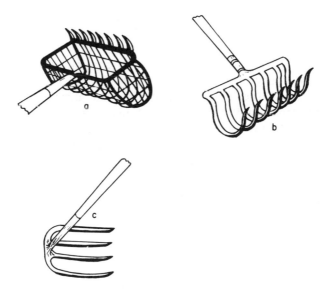

Figure 2.13 Clam rakes. (a/b) Hard clam rakes. (c) Soft shell clam hoe or fork. (*Wheaton and Lawson, 1985, p. 101*)

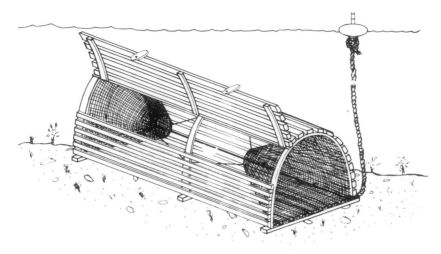

Figure 2.14   A New England lobster pot. (*Wheaton and Lawson, 1985, p. 85*)

FIXED VERSUS MOBILE GEAR

The fishing industry makes a distinction between fixed and mobile gear. Mobile gear is supposed to be maneuvered to avoid contact with any fixed gear. If available fishing space is limited, usually a reflection of the fishes' biologies, significant conflicts can arise between competing fishermen.

Unfortunately, disputes such as those between fishermen using different gear are reminiscent of the range wars between American cattle and sheep farmers of the 19th century. Gear conflicts have often led to violence. These intra-industry struggles also prevent the fishing industry from working together on some of the more significant issues that affect the entire industry. Prior to the extension of coastal jurisdiction, one of the major complaints of American fishermen against foreign fleets was that the large foreign draggers tore up a lot of fixed American gear (e.g., lobster traps); they alleged that no effort was made to identify and pay back the owners of the fixed equipment. With the new jurisdiction, the rules for fixed-versus-mobile gear are more clearly defined and enforceable, and are in fact better enforced, particularly on the foreign fleets. The government's commitment to enforcing these laws is extremely important to fishermen.

REFERENCES

Wheaton, Fredrick W., and Thomas B. Lawson. 1985. *Processing Aquatic Food Products.* New York: Wiley.
Yamaha Motor Co. Ltd. 1986. *Yamaha Fishery Journal Composite.*

3

Onboard Handling and Processing of Fish

Once the fish are on board the ship, the crew must process and stow the catch in preparation for shoreside unloading, further processing, marketing, and distribution. They must work quickly and efficiently in spite of any problems such as faulty equipment or inclement weather.

The first step is to sort the catch by species, since the most valuable species are generally processed first; this sorting sometimes also includes size sorting.

In a few special cases, for example, for the Japanese market, the fish may be killed actively by pithing, that is, spiking of the brain. Called *ike-jime*, the process is believed to yield a higher-quality fish. *Ike-jime* slows down the fish's postmortem biochemistry (i.e., the changes after death); for example, it slows down postmortem pH changes and the rate of ATP breakdown.

For some fish species or sizes, there is no ready market or the work involved in preparation is inordinate for the retail price; fishermen throw these fish back overboard. If these fish are believed to compete for food with the primary local species that are sought, the undesirable fish may well be killed onboard with a quick knife cut before they are tossed back. This practice constitutes a tremendous loss of food in light of the significant resources that have already been invested in getting the fish to this point. As much as 90% of the catch may be tossed back, for example, in some of the Great Lakes fisheries, as well as with shrimp by-catches around the world. Clearly, more effort to develop new and better uses for the by-catch (or underutilized fish) could help feed a hungry world.

On the other hand, regulations sometimes require the crew to avoid higher-value species, for example, when only a specific percentage or amount of incidental catch of a more expensive species is allowed and must be reported to the appropriate governmental agency. In some cases, techniques can be developed for minimizing the incidental catch, such as of crabs when fishing for sole off Alaska. When the trawl doors move through the water, sole tend to move off the ground, while crabs tend to bury themselves. Thus if the "footropes," that is, the lower ropes of the net, are loosened slightly, they will pass over the crabs but still catch the sole. Another example from the same locale is the need to avoid halibut. These fish prefer somewhat higher/harder ground that can be mapped and avoided.

If the by-catch becomes too large, an entire fishery may be closed down. This happened in Alaska recently, where the halibut by-catch led to the closure of the bottom fishery (i.e., cod, flatfish, and pollock). Many of the factory trawlers have been unable to fish. A consequence has been a switch to modified crab-traps to catch these fish.

INITIAL HANDLING

Bleeding and Gutting. Depending on the customs of the area and the species, the fish may be bled on board (generally done to higher-value, white-fleshed species). For example, a fish like dogfish really should be bled to remove some of the urea in the blood, which otherwise becomes available for ammonia build-up. This is not always done because of the desire for a redder fish in some countries (e.g., Germany); the remaining blood in the white flesh gives a red tone. Bleeding is usually accomplished by cutting the arteries just behind the gills and in front of the heart. The fish are then allowed to bleed either in air or in water for about 20 minutes. Water is generally considered to be the better choice even though it requires a special "bleed tank" and periodic changes of water. Sharks and flatfish are bled by having their tails "bobbed," that is, cut off.

Following bleeding, the fish may receive a final wash. A suggested technique for decreasing bacterial growth in fish is the use of a very high-pressure washing system; this system might also extend the fish's shelf-life. Blanching, that is, dipping the fish in water that is close to boiling for 2 to 3 seconds, has also been suggested.

Following bleeding, the fish may be eviscerated (gutted). This is generally done manually: the fisherman cuts from either the fish's gill cleft to the anus or from the anus to the gills, and then tears out the guts ("ripping"). Equipment for gutting fish is usually effective for particular sizes of fish. For example, the Danish Jutland machine can offer a reject rate as low as 2%. Two machines are usually needed, one for fish under 55 cm (21 inches) and one for fish over 55 cm. A different approach is used to mechanically gut flatfish. A "cookie cutter" hole is stamped out in the gut region carrying the guts with it.

The fish must still be alive for proper bleeding, and therefore cannot be left untreated onboard for too long. In many fisheries the fish are gutted and bled at the same time in a one-step operation. This may lead to less bleed-out. With manual gutting, part of the reason for an incomplete bleed-out seems to be the failure of the fisherman to properly (deeply) cut the main artery, located just below the backbone. Machine-gutting always cuts this artery and may do a better job at bleeding, but it may not get the belly cavity quite as clean.

Species like redfish (sold in the United States as ocean perch) are relatively small and have very spiny fins which would make it dangerous for the fishermen to gut them at sea. These species are generally filleted onshore while still ungutted. The resulting lower yield is more than made up for financially by the labor savings in not gutting these fish.

Interestingly, flounder-like species, that is, flatfish, are not generally gutted in the United States. The bulk of these species are eventually sold as fillets prepared directly from the whole fish. In certain other countries, these fish are sold whole and served with the head still on. For example, in New Zealand, it is one of the few inshore species of fish that is gutted at sea.*

There are many reasons for gutting fish at sea. Removal of a fish's viscera eliminates the potential contamination of the edible flesh with various hydrolytic (digestive) enzymes found in these organs. It also prevents the inoculation of the flesh by the large microbial population found in the intestinal track. At the time of catch the flesh itself is probably sterile. Other potentially damaging materials such as the bile from the gall bladder may also be released if the ungutted fish are mishandled. The bile is an alkaline, greenish liquid, which can cause a visible discoloration referred to as "belly burn" that gives fish a bitter flavor.

Quality Issues.   If fish have been feeding, that is, if they have eaten recently, the enzyme levels in the intestinal tract may be high; their bellies may also be extended beyond the normal contour of the fish, making them more easily subject to bruising and the resulting belly burn. Following breakage, worms normally found in the intestinal tract may migrate into the edible flesh.

Fish gills may also be removed for certain species and particular markets, for example, the "Princess" cut for salmon destined for Japan. There is no evidence that removing the gills affects the quality of the fish during storage, but, as will be seen in Chapter 6, their removal makes it much more difficult to determine the actual quality of the retail fish. The gills probably do have a fairly high microbial level but, to date, research has not confirmed that this affects the flesh in any way. The removal of fish gills during gutting remains controversial. For example, the new Portland, Maine, auction requires gilling during the warmer months. On the other hand, results at the Torry Research Station in Aberdeen, Scotland, suggest that the gills are one of the best indicators of retail fish quality and that their presence does not increase the rate of fish spoilage.

Good workmanship, that is, the care with which the fish are handled, and sanitation produce a better product. Pitchforks, gaffing hooks (i.e., a wooden

---

* The term "inshore" refers to those fish that are caught close enough to shore that they are generally caught by "day boats," that is, boats that only go out fishing for the day. "Trip" boats stay out overnight, can therefore go much further, and will often go "offshore."

stick with a nail at the end that is used to hold and move a fish), and other devices that penetrate the fish's flesh are a source of contamination and should be discouraged at all times; the wounds formed are often the sites of further spoilage. If a gaffing hook must be used, care should be taken to clean it regularly and only to use it on fish heads that will eventually be discarded. A washable one-piece construction of stainless steel is much superior to a piece of wood with a rusty nail driven through it.

Fish should not be thrown around. Neither should they be handled by the tail; since the tail is often unable to support the weight of the fish, this practice can break the fish's backbone. Rough handling can cause internal bruising and bleeding that becomes visible either when the fish are cut into fillets or steaks or when they are eaten by the consumer. If a steak or fillet is bruised, it must be rejected or at least trimmed, causing financial losses in terms of labor and yield.

The reader must remember that although we are discussing various aspects of onboard handling from the point of view of improving fish quality, the incorporation of these methods by the industry will probably center on economics. An improved yield or a higher price for quality is the strongest motivation for fishermen and processors.

A new system of dockside grading is being developed for Canada's Atlantic fleet. The regulations include standards for bruises and fillet color, which processors believe to be the major cause for devaluation of fish upon unloading. Research at the Canadian Department of Fisheries and Ocean's Newfoundland laboratory suggests that many fish cannot meet this standard at the time of catch because of the rough handling received during the catching/landing process, especially with large trawls.

Onboard Storage.   If the fish is not going to be gutted or further processed onboard, then it must be stored appropriately for the duration of the trip. The average fresh fish trip is probably 5 to 6 days long and may be as long as 12 days. Even under the best of circumstance, the fish is expected to lose weight because of "drip," the whitish liquid that comes out of fish over time. The fish must be "iced down": the fish should not be stored without ice or some form of cooling; even day boats must eliminate the practice of leaving fish sitting on an open deck.

The least satisfactory way to ice fish is to simply put some ice down on the bottom of the hold (cargo area) and pile up the fish, sometimes covering the fish again with some ice on top. Two problems with this system should be obvious. First, many of the fish are too far away from the ice to be cooled efficiently. The thermal conductivity (i.e., the ability of a material to transfer heat through itself) of the fish mass becomes a major barrier to cooling. With some metabolic activity still taking place within the fish flesh, the mass of fish in the middle might even heat up. The second problem is that the weight of the fish pile crushes the fish on the bottom, leading to a loss of weight (yield) as well as other physical damage. For example, if a short, 3-foot-deep pile of haddock is stored for 2 weeks, it has been reported that the bottom fish lose 15% of their weight as compared to a more normal weight loss of 3 to 8% (due solely to biochemical changes that cause a loss of water retention, leading to drip).

OPTIMIZING THE ICING PROCEDURE

The more intimately the fish are mixed with the ice, the better the cooling. The melting of ice creates a slow water flow that continuously cools and washes the fish. This is best done by alternating the layering of ice and fish.

Ice should be applied directly to the belly cavity of large fish. This area has a high initial bacterial load and is therefore more prone to spoilage. The belly should be placed downward so it does not trap accumulating melting water that would encourage bacterial growth.

Boxing. Boxing or short-shelving, that is, the use of many shelves so that each shelf has only one layer of fish, may be used to avoid crushing. However, if either technique is abused, for example, by piling the fish to a higher level than the box top or shelf level, then the potential advantages are completely lost.

The nature of the boxes used varies in different places. The same type of boxes can often be used for onshore handling of fish. Boxing is preferred by the shoreside processor over short shelving because it saves effort during unloading, subsequent dockside handling, and, therefore, abuse. It is more labor-intensive than short-shelving and does not always take advantage of the available space aboard ship. Older boats were not designed with rectangular fish holds. Efforts to introduce boxing at sea in New England have had very limited success because of the additional labor involved and the lower total trip capacity onboard. Fishermen feel that they must get a premium for these fish, which are usually of higher quality, but the marketing system in most New England ports makes such a premium difficult to obtain. Some success with boxing has been reported from Maine.

The wooden box is probably the cheapest and oldest type of fish storage box used. It is difficult to clean and may occasionally lead to splinters in the product. However, it is relatively nonslipping. In some ports metal boxes or "kits" (a tapered barrel-like device) are used instead of wood. Many other boats and ports have moved to plastic boxes. These often have special designs to allow safe stacking on board ship, that is, special top and bottom surfaces that prevent load shifting.

Some boxes have a nesting feature that minimizes their total empty volume. There is usually a penalty for this feature in the form of a decreased internal capacity per box when full; because they are tapered the boxes are not space-filling when full. The boxes may also be designed so that the meltwater cannot go through the boxes below, so preventing cross-contamination. Some companies advertise that the water's flowing through to the lower boxes better utilizes the cooling and washing power of the ice, but this is a questionable practice at best.

Improving Fish Handling.   There have been a number of recent developments to improve onboard handling of fish. In Newfoundland, Canadian fishermen have been experimenting with various types of bulk handling containers. One of the systems developed, "Contrawl," involves large overhead rails that automatically load, move, and off-load the large half-meter-deep (18-inch) containers.

The Food and Agriculture Organization of the United Nations (FAO) has published a study including necessary planning and engineering data for improving fish handling systems, especially by improving the understanding and use of fish handling containers (Wood and Cole 1989).

In many ports the availability of ice is limited and the cost of ice in proportion to the value of the fish may be quite high. In these cases, efforts to get fish properly iced depend on having the necessary infrastructure that can provide the ice and pay some premium for properly handled fish.

Other fishermen use more sophisticated methods of handling fish, for example, chilled seawater (CSW: ice is added to seawater) or refrigerated seawater (RSW: a mechanically cooled seawater). With any system that includes circulating water it is important to insure that the system does not become clogged with debris. Some sort of strainer must be used.

Sometimes a "champagne" system may be incorporated to ensure good mixing of the ice and the water by bubbling air (or on a few more sophisticated boats, carbon dioxide) through the cooling liquid containing the fish. However, if the air is warm, it will be adding heat to the system. Also, the bubbling of air rather than carbon dioxide can encourage rancidity to develop in the fish by providing a greater amount of oxygen. On the other hand, carbon dioxide acidifies the water, increasing salt uptake and lowering rancidity development. These systems are often used for smaller, fattier fish such as roe herring (that is, herring still containing roe, popular in Japan).

The physics of these more complex systems is different from that of the standard icing procedure. The intimate contact and heat-transfer properties of liquids allows for a more rapid cooling than occurs with ice. Ice generally has a number of air pockets which serve to slow down the heat transfer since air is an insulator. Unfortunately, the equipment for these more complex systems takes up a lot of room on the boat and may be expensive. If a large load of fish is added to a tank at one time, the CSW system may cool the system somewhat quicker than a mechanically cooled refrigeration system because of its faster response rate; that is, the extra ice simply melts faster and, in melting, quickly removes more latent energy of melting. The heat requirement in the change of state from solid to liquid causes cooling.

D. Amos, at the University of Rhode Island, has calculated some useful relationships with respect to the amount of fish that a boat can hold using different handling techniques. For example, a 75-foot boat might be expected to have a fish hold of around 3,531 cubic feet, while a 118-foot boat would have a fish hold of around 8,123 cubic feet. Based on these two hold sizes, Table 3.1 indicates the amount of fish that can be carried using various types of onboard storage systems and species. Notice that some of the systems require a great deal more space than others.

Fish can only be left in liquid saltwater systems for a limited amount of time before becoming soggy or absorbing too much salt. These systems generally work best if the fish can be off-loaded, or otherwise removed from the system, in 3 to 5 days. At the end of the trip, if the tank is so designed, it may be drained and directly off-loaded, saving handling and damage. This, of course, is also quicker than many other methods.

The use of carbon dioxide in these systems is based on the general toxic effect it has on bacteria. Other chemical additives, such as potassium sorbate, have also been tested in CSW/RSW systems as antibacterial agents. Antibacterials are discussed in Chapter 5.

Some fish, such as large tuna, require special handling if they are to receive a top grade in the marketplace. For tuna, this includes being sold in Japan for use as raw fish—sashimi. Tunas have such an active metabolism that their body temperature may run significantly above the ambient sea temperature. In at least one case, a West Coast albacore tuna was measured

Table 3.1   Fish Hold Capacities

	Boat hold capacity (tons)	
	75 ft	118 ft
Bulk herring	70.0	160.0
Bulk fish	46.2	107.0
Shelf fish	24.0	54.0
Boxed, 100 lb	29.0	67.5
Boxed, 55 lb	38.6	90.0
Frozen	50.0	115.0
CSW or RSW tanks	40.0	92.0

Courtesy of *National Fisherman* and Duncan Amos.

to be 16°C (30°F) above the ambient water temperature. Once killed, if ice is limited, the fish should be left in the water for about an hour to allow time for the temperature to come down to ambient. They should then be landed and immediately gutted and iced, including placing of ice inside the belly cavity.

ICE

Ice is extremely important to the fishing industry. Its many properties are therefore worthy of specific discussion.

Types of Ice and Their Qualities.   A number of different types of commercial ice machines are currently available. Some of these make what is called "wet ice" and others make what is called "dry ice." The latter is not to be confused with the use of the term "dry ice" for frozen carbon dioxide, which is much colder than ice. The difference between these two types of ice has to do with the process used to remove the newly formed ice from the ice-making equipment. If the ice-machine surface is heated, usually with hot gas, so that free water exists on the surface of the ice, then the product is wet ice. If stored at temperatures much below the freezing point of water, wet ice resolidifies and the individual pieces rebind. Dry ice is usually removed from the equipment by some mechanical action other than heat. If it is not allowed to melt, dry ice does not have the problem of resolidification and rebinding.

Ice generally does not flow and—particularly with wet ice—care must be taken to prevent bridging during storage; the emptying of ice below the "ice bridge" leaves a continuous bulk of ice suspended in air. A person going into the storage area to break up this ice bridge must exercise due care for his or her safety; wearing of clean clothes helps prevent contamination of the ice.

Ice can be made from pure water or from water with various additives, for example, potassium sorbate or salt. The characteristics of these treated ices may be somewhat different. Saltwater ice tends to be particularly slushy. If large blocks of chemical-containing ice are being frozen, it is important to insure that the added chemical remains evenly distributed. This is not always easy: the freezing process generally starts by freezing of pure water in contact with the cold surface; the salt in the bulk liquid becomes concentrated in the liquid phase until the eutectic point of the salt-water mixture is reached, at which point the entire system freezes.

If a solution contains a higher concentration of salt than the eutectic concentration, the solution actually freezes at a temperature higher than the eutectic temperature.*

---

* The complete freezing of all the water that is "free" or unbound in foods may not occur until about −30°C (−22°F).

Of course, ices with salt in them have a lower melting point than pure water. For example, saltwater ice melts at about −1°C (30°F).

Freezing point depression is a colligative property, that is, a property that depends on the actual number of molecules present in solution; thus a 1 molar solution of an unionized compound lowers the freezing point 1.8°C (3.24°F); a 0.5 molar solution of a monovalent–monovalent fully ionizable salt like NaCl will form 1 mole of ions and will, therefore, have the same freezing point depression. A more elaborate chemical equilibrium calculation is needed with partially ionizable compounds to determine the amount of each form of the compound present in solution under the working conditions. Thus the freezing point depression of some solutions may also depend on their pH.

Ice becomes opaque with the addition of additives like salt. Since diluted seawater ice has more cracks than regular concentrated seawater, and rapid freezing to −40°C (−40°F) causes greater opacity and more cracking than freezing at −20°C (−4°F), the ice surrounding some fish looks quite opaque. Fresh water is therefore recommended for freezing ice that is used in consumer presentations.

Role of Ice in the Fish Industry. The proper use of ice requires an understanding of its role in fish quality maintenance. Ice serves three essential functions: cooling, washing, and moisturizing. In addition, if the ice is not clean, a surprisingly common occurrence, the ice can serve as a source of bacteria, particularly psychrotrophic (cold-tolerant) and psychrophilic (cold-loving) bacteria that enjoy life on ice. The first function, cooling, requires the maximum contact of ice with the product to be cooled for the best heat transfer properties. Air, a natural insulator, is to be avoided. Oftentimes, a product may seem to be properly buried in ice but, because of the equivalent of the bridging phenomenon that may occur following a small amount of melting, an air layer may form between the ice and the fish. The motion of a boat generally alleviates this problem by collapsing these air gaps; air gaps are therefore more likely to occur with fish stored in a cooler in the processing plant.

The washing function of the melting ice leads to the removal of surface bacteria. It is generally not desirable to have the dirty meltwater go through the entire stack of fish. For melting to occur at a meaningful rate, it is necessary to set the holding cooler's temperature slightly above the freezing point of water, that is, 2–3°C (35–37°F). The rate of loss of ice must be calculated carefully so that the fish are never held without ice.

The moisturizing function simply refers to the need to keep the fish's surface wet so that it glistens when viewed by the consumer. The humidity to do this is supplied by the melting ice water. With mechanical refrigeration, it is important to insure that the humidity remains high. If the fish are properly cooled and cleaned before being stored, then the main role of the ice should be to maintain temperature and humidity.

It takes about 0.44 to 0.50 calories to heat 1 gram of ice by 1 degree Celsius. At 0°C, where the phase transition of ice to water occurs, a great deal of heat is needed to change the ice to water (the latent heat of the ice–water transition is approximately 79 calories per gram). The water can then be heated to higher temperatures, requiring approximately 1 cal/g/°C. The normal definition of a calorie is the heat necessary to raise 1 gram of water by 1 degree Celsius between 14°C and 15°C. Thus, the ice serves best as a sink of energy for heat removal at the phase transition. Keeping ice colder than 0°C provides little extra cooling benefit. This is why stored ice need only be kept slightly below 0°C to prevent melting.

Where does the heat come from to melt the ice? From the fish, of course, but also from the surroundings: the air, the boat, the fish box, and so on. Once the fish are cooled to 0°C, the only sources of heat are the surroundings. It is therefore a good practice to put extra ice on the top, bottom, and sides of the fish storage area. In a well-designed fishing boat the fish hold is as far away as possible from major heat sources such as the engine.

Selecting the Right Ice.   This information on the physical chemistry of ice helps answer the often asked question, "Which ice is best?" All ices are essentially equal in cooling power on a weight basis. However, on a volume basis there are differences between the different types of ice.

The smaller the ice pieces, the more surface-area contact there is with the fish; this increases the possibility of a more rapid melt (i.e., better cooling). This is desirable initially to speed cooling but can later lead to a too-rapid melt of the ice. Large pieces of ice make poorer contact with the fish surfaces, and if they are sharp they may also cause mechanical damage and bruising.

How much ice is needed? The general rule of thumb is about 1 kg of ice for each 2 kg of fish in temperate climates, with slightly more in summer, and slightly less in winter. One kilogram of ice per kilogram of fish is best for fish caught in tropical waters. Important related factors include the total amount of ice needed, the space in the hold that it takes up, and its cost; for example, in the United States ice costs about 2 to 3 cents per pound. The benefits of using ice with respect to fish quality may not show up until much later in the marketing chain, making it harder to convince fishermen of the importance of icing on fish quality.

Table 3.2 gives additional data on the use of ice for cooling fillets, both thermodynamic and kinetic. It should also be remembered that ice expands on freezing.

Because of the limited amount of heat that can flow through a fish in a unit of time (i.e., the fish's thermal conductivity), the process of cooling takes

Table 3.2 Use of Ice

Ice Required to Cool 1 kg Fillets to 0°C

Initial Temperature (°C)	Weight of Ice (kg)
20	0.25
15	0.19
10	0.13
5	0.06

Cooling Time Related to Distance (Wet Fillets)

Distance[a] (cm)	Time (hours) to Cool to	
	5°C	2°C
1.5	<1	4
2.5	2	18
5.0	8	>24
7.5	18	
10.0	>24	

Taken from Aitken et al. (1982)—Crown copyright.
[a] From surface to cooling point.

time; the thicker the fish to be cooled, the longer it takes to cool that fish. Although it generally only takes 1 kg of ice to cool 7 kg of fish from 13°C (a relatively warm temperature for temperate-water fish) to 0°C, the extra ice that is recommended above 1 kg is suggested as protection against the heat input from the environment.

Calculations of ice needs are also required in the retail store. The publication *Progressive Grocer* recommends that the retail display case for fish have a stainless steel pan capable of holding at least 8 inches of ice. This should be used as the base for the display. The refrigeration unit should be able to deliver up to 280 BTU per hour for each running foot of the display case when operating against a 100°F (38°C) ambient temperature. Others feel that the ice should only be there for decorative purposes and to supply some moisture, but that the temperature should be maintained by the refrigeration system.

Some of the relationships with respect to heat transfer and ice can be described mathematically and are discussed.

Heat Transfer

Heat transfer from the inside to the outside of a box is given by

$$q_{transferred} \text{ (kcal/day)} = A \times U \times (T_i - T_e)$$

where

$q_{transferred}$ = heat transferred in unit time
A = area of heat transference (m^2)
U = overall heat transfer coefficient of the box material (kcal/(day m^2 °C))
T_i = temperature in the box (°C)—0°C?
T_e = temperature outside the box (°C)

To maintain high fish quality, this heat flow must equal the heat obtained from the melting of ice:

$$q_{fusion} \text{ (kcal/day)} = H_f \times \frac{dM_i}{dt}$$

where

q_{fusion} = the total heat of fusion supplied in unit time
H_f = latent heat of fusion (80 kcal/kg) for water/ice
M_i = mass of ice (kg)
t = time in days

In other words, dM_i/dt is the rate of loss of weight of ice. At equilibrium, that is, the steady state, $q_{transferred}$ equals q_{fusion}. The equations allow us to calculate the heat transfer coefficient (U) for a box of known surface area if the weight of the ice lost in the box is monitored along with the temperatures on the inside and outside. Once this property is known for a particular material, we can determine the rate of ice melt (kilograms of ice per day), that is, the slope of a plot of mass of ice versus time for any container made of the same material. Thus, ice needs can be predicted ahead of time. The theoretical values correlate rather well with experimental values.

All this information helps us better determine the amount of ice needed for particular conditions—not forgetting, of course, the additional ice needed to bring the product to 0°C. These data allow for good rationalization of ice usage.

The specific heat capacity of fish is normally calculated from the composition of the fish:

$$C_{pf} = 0.5X_f + 0.3X_s + 1.0X_m$$

where

C_{pf} = specific heat capacity (C_p) of the fish (kcal/(kg °C))
X_f = mass fraction of lipids (kg/kg)
X_s = mass fraction of solids (kg/kg)
X_m = mass fraction of water (kg/kg)

The heat capacity for lean fish is about 0.8 kcal/(kg °C); for medium fatty fish it is about 0.78; and for fatty fish it is about 0.75. For water it is about 1.0. Therefore, the amount of ice needed to cool the fish (assuming it is cooled to 0°C) is

$$\frac{C_{pf} \times T_f}{H_f \times M_f}$$

where

T_f = temperature of the fish before cooling (°C)
M_f = mass of fish (kg)

The trick is to balance the ice-to-fish ratio so that the volume of any box is fully used, particularly to maximize the amount of fish therein! Again, the key components are the amount of ice to cool the fish and the amount of ice lost per day from a box, which depends on ambient temperature. It is sometimes also helpful to know the specific volume of the fish stowed (cm³/kg) and the specific volume of the ice; for example, 1,274 cm³/kg for the fish (barracuda) and 1,371 cm³/kg for flake ice.

The best economic ratio of ice to fish can be calculated when we also know the cost of ice, the average ambient temperature, and local wage costs. In some countries if often pays to maximize ice usage, while in other countries it clearly pays to maximize the efficiency of labor.

OFF-LOADING

It is easy to remove fish from the boat improperly; unfortunately, it is also very common. Before the unloading process begins, the hold itself should be checked for safety. If fish have begun to spoil, the hold may contain high levels of carbon dioxide, hydrogen sulfide, or other noxious compounds. If this is the case, the hold hatches should be left open for some time before anyone enters.

In many ports, the fishing crew does not unload the boat; rather, a shoreside workforce (lumpers) comes aboard for this task. Winches are often used to move boxes of fish; sometimes a simple wire or rope basket is filled with fish that can then be brought from the boat to the dock. The fish are often shoveled or pitchforked (even worse!) into these baskets, then dumped into containers onshore. Depending on the intended further handling, that is, whether the fish will be held for a time or processed immediately, fresh ice may be added at this point. In many plants, as the fish are moved through the system, a great deal of further "dumping" occurs. Some boats use mechanical elevators and the like to get the fish out of the hold, but there is still a great deal of hand shoveling. If these hold-emptying devices are overfilled, fish sometimes fall a significant height and are bruised.

Better systems of off-loading the fish are being developed. Some of the newer systems pump the fish. Several require that the hole be flooded (wet-pumps), while others do not (dry-pumps). The wet-pump systems are also used to "brail" fish such as herring and menhaden from seine nets. They often work better on live fish because of the fish's greater flexibility. Because the fish have almost no flexibility during *rigor*, pumped fish are often badly damaged. Recent advances have improved pump design such that plants handling traditional groundfish-sized fish this way are not sacrificing quality.

Once ashore, some plants move fish between points by fluming the fish in very gentle flows of water. Others use various conveyors and reboxing schemes. If fish are being moved whole to other markets, they may simply be washed and repacked. The wash water is often nothing but a tank of dirty water! The 50- or 100-lb "wet-lock" box is probably the standard shipping box used for whole, or dressed, fish. These boxes are made of heavy-duty corrugated cardboard coated with an appropriate wax on both sides to make them moistureproof. The fish are sometimes iced only at the top and bottom. The discerning customer should seek more ice between fish layers.

REFERENCES

Aitken, A., I. M. Mackie, J. H. Merritt, and M. L. Windsor. 1982. *Fish Handling and Processing.* Edinburgh, Scotland: Her Majesty's Stationery Office.
Wood, C. D., and R. C. Cole. 1989. Small insulated fish containers. *FAO Fisheries Circular No. 824.* Rome: FAO.

THE STORES
INSTITUTE OF AQUACULTURE
PATHFOOT BUILDING
UNIVERSITY OF STIRLING
STIRLING FK9 4LA

4

Onshore Handling

There are strong sociological trends that have shaped the fishing industry. (1) Many decisions have been based on tradition and other nonscientific rationales. These may be more reliable than pure guesswork—but then again, they may not be. (2) The new alliance of technologists and "fishing folk" must be based on mutual respect. Disagreements are inevitable, but the fishing industry can only thrive if the "long journey" is started in a spirit of cooperation. (3) The processing of fish is a very practical matter. There are many steps to the dance, and many dancers on the floor. The official authorities of each state have not been the only parties providing security and other necessary services to the onshore fish handlers.

Fish can be processed in many ways depending on the species, location, personnel, and equipment resources. If fish are boxed at sea, the full boxes can be off-loaded immediately. However, since current equipment for accurately weighing fish at sea is limited, many markets require that these boxes be reweighed when they reach shore. If the fish are stowed onboard in bulk, they may be put into containers at the point of unloading.

THE FISH MARKET

Some processing companies avoid price negotiations by running their own boats. In these cases, it is difficult but important to emphasize the incentives for quality rather than just quantity.

Regardless of the method of off-loading used, the fish can be sold by the catcher to the first buyer in one of several ways. This usually occurs close to the point of landing.

The major fish markets overseas are London's Billingsgate, Paris' Rungis, and Tokyo's Tsukiji. In the United States, the major markets are probably New Bedford, Boston, New York, New Orleans, Terminal Island (Los Angeles), and Seattle.

Fish Auctions.   In the United States fish may be sold on consignment, by a prearranged contract with or without a set price, or by the use of some form of auction system. The auction may be arranged so that the whole boat load is sold as one lot, that is, the entire catch aboard the boat must be purchased with a single bid by one company, even if it does not want all the fish on the boat. Often used in New Bedford, Massachusetts, this method leads to a lot of secondary handling as the purchaser of the boatload attempts to redirect to other buyers the fish he has not already sold or does not wish to process.

The crew's usual responsibility in the auction procedure is to accurately describe the catch, or at least the type and species aboard. The only control of fish quality is the buyer's general knowledge about how the captain and crew handle fish and the amount of time the boat was out at sea. The final payment is on a per-pound basis and is based on the actual weight of fish that is landed. In the New Bedford auction, the price per pound of all fish onboard is posted, and the price for any one of the species listed can be increased in increments of 0.1 cents during the auction. To overbid the previous bidder for the entire boatload, the new bidder needs to increase the price for only one of the species aboard. In terms of the amount of the overbid, it makes a big difference whether the bidder raises the price on 10 lb of a minor fish or on 7,000 lb of the major species aboard.

These proceedings are very different from the method used in Boston, for example, where a specific part of the catch can be bought. Nevertheless, buyers still do not actually see the fish they are purchasing until after the auction is over.

Portland, Maine, has set up a European-style display auction in which the fish are laid out in boxes and purchasers can buy by the individual lot. The fish are displayed without ice for examination purposes, so the temperature of the display hall must be carefully controlled. Unless the fish are boxed at sea, this also means additional handling as the fish must be boxed for the final user. The European-style auction procedure is meant to encourage fishermen to care about quality, since buyers can really see what they are purchasing. The system also allows smaller-lot purchases, which may bring more buyers to the market. Current plans call for a minimum size of twelve 100-lb boxes making up a pallet to a maximum of twelve such pallets as one lot. The Portland auction also takes fish brought in "over-the-road," that is, brought in by truck from elsewhere. The extra work involved in shoreside boxing and smaller-lot sales must translate into greater profit for the fisherman, who is now working harder to achieve higher-quality fish. The exchange's estimate of the cost for culling/weighing (at the rate of 45,000 lb/hr) is six cents, with the buyer and seller each being charged three cents. At this time it is not clear whether a display auction will find a permanent place in the United States fish marketing system. The Portland auction ran into serious financial problems in 1988, but after undergoing a significant restructuring it seems to be flourishing.

In Aberdeen, Scotland, the fish are put out in wooden 6-stone (84-lb; that is, 1 stone equals 14 lb) boxes. The auctioneer takes the bids. The winner then places his marker, a paper strip with the company name on it, on those boxes he wishes to purchase. Thereafter, anyone who wishes to purchase any remaining fish at that price may put his or her marker on other boxes. The bidding is later reopened until all of the boxes are sold. Because the market participates in the European Economic Community's Common Fisheries Policy, it is covered by a withdrawal price bottom (see Chapter 16). Fish that does not sell for at least a certain minimum price is bought by the government and given to qualified charity organizations.

Variations of the auction system include the "Dutch clock" auction, in which the opening price is clearly too high and the clock works down from there. The first bidder in the Dutch Clock auction gets first crack at the fish. Because the product is sold on the first bid, it is felt that this system makes it more difficult for buyers to collude. The Dutch clock was the auction system being proposed for use by the New York and New Jersey Port Authority's "Fishport" at Erie Basin in Brooklyn. This $27 million fish landing, processing, marketing, and storage project was supposed to include 27,000 ft^2 of auction space with the capability of handling 250,000 pounds of fish at a time. However, after spending a great deal of money on the project, the Port Authority has decided to close Fishport. Apparently, its landings for the first year only equalled the volume handled by the Fulton Fish Market in two days! The failure reflects the difficulty of a large agency's participating in the fishing industry. The people making the decisions did not really understand the fishing industry and all decisions had to go through too many layers of bureaucracy.

Electronic Marketing.   Electronic marketing via computer would permit buyers and sellers to interact at a distance. Although several studies have focused on the exciting possibilities, more attention must be directed to the problems concerning appropriate descriptions of products for sale. It is difficult enough to describe fish quality at the time of sale in a standardized way; this becomes more significant when done "long distance." And then, of course, there is the problem of trying to use current information to predict the fish's quality at the time of delivery. Are predictions to be based on the expectation of proper handling? What if unexpected problems arise in the time (as much as a few days) between sale and delivery? The amount of time between sale and delivery is important. So is the method used for transporting the fish. Researchers are trying to develop more objective and reliable methods of evaluating and predicting fish quality (Chapter 6). We would suggest as an immediate possibility the use of the Torry Freshness Scoring System (Chapter 6).

In some ports, fish are simply packed into boxes with ice and sent off to the market on consignment. The broker at the market sells the fish at market price, takes his or her commission, and remits the remainder back to the fishermen. The fishermen do not learn the price they will receive for their fish until some time after they have landed the fish. This system is widely used in the New York area, especially the Fulton Fish Market, the city's wholesale fish market. A drawback of the system is that fishermen who are not in close touch with the market begin to assume that they are being "taken" and become resentful. The possibility is heightened because of their assumption that prices at the Fulton Fish Market are "fixed by organized crime." However, in any market, if a price moves significantly away from the true price, alternate marketing channels are sought and developed. In fact, although the Fulton Market remains the largest and most important market in the Eastern United States, recent years have seen the development of many more alternate arrangements by fish companies.

Fish prices in key markets are widely circulated through a computer service linked to the National Marine Fisheries Service (NMFS). The prices officially reported at these markets serve as the basis for further trading. Unfortunately, these prices are subject to manipulation. Since the amount of trading taking place in some of these public markets is limited, an individual can "adjust" these prices to enhance profit on private sales. For example, in New Bedford for many years, only the eight unionized processing plants could bid for the fish on the auction and the auction was controlled by the union! During a recent strike, a new auction was created. Since the strike the new auction system seems to be the main auction in New Bedford.

In Boston, where the volume on the market is about 6% of what it was in 1936, the careful timing of the arrival of shipments from Canada can have a distinct effect on the price on a particular day. Boston prices, unlike New Bedford's, are based on a per-ton price; current prices for fish may be based on 40,000 to 50,000 lb of fish involving only three key buyers. As a frame of reference, a single tractor-trailer holds about 40,000 lb. The New York market has often had a fixed price for major species, as discussed above. Thus, the value of fish on a particular day has often been a very controversial issue, with the result that fisherman generally feel that they are being cheated. The system in New York actually gives the buyer an incentive to know about fish quality—the best buyers will purchase the best fish at the "fixed" price.

PRE-FILLETING HANDLING

Scaling.   If the fish will not be skinned, for example, redfish (ocean perch) and sometimes haddock, then scaling may be required prior to filleting.

Because haddock is worth more than cod, the skin proves that it is really haddock. The most common method of scaling is to use a tumbler with a rough surface and lots of water spray. Unfortunately, such a method subjects the fish to a lot of mechanical abuse. A second method of scaling—using a special machine with brushes—is much faster but costs more.

The scales obtained from this process are generally discarded. Those from a fish like herring may be saved for pearl essence, used to manufacture nail polish. The high level of guanido compounds gives the scales their special sparkle. Research at Cornell (Welsh and Zall 1979) suggests that the scales can also be used in waste water treatment of edible food wastes (Chapter 10).

Beheading.   Some of the machines used for filleting fish require that the head be removed first. Special machines are available to "nob" fish, that is, head and gut them on the same machine. Some are designed so that as the head is pulled off, the guts are removed simultaneously.

FILLETING

In most primary markets the fish is sold in the various "whole" forms including totally "whole," gutted, gutted and gilled (Princess cut), or headed and gutted (H&G, nobbed). In many secondary markets, for example, the Fulton Fish Market, the fish may also be sold as fillets, and sometimes as steaks; these may be, for example, "over-the-road" fillets, which have been cut elsewhere, or fillets that are actually cut at the market. Many of the "over-the-road" fillets coming into New York City are processed in the disparate filleting houses found along the coast, particularly in New England. Some are small shacks with a couple of people working with knives—indoors or outdoors! Others are modern food plants with a large, trained work force producing both manually cut and machine-cut fillets. Thus, the buyer must either be very good at identifying quality, know a lot about all of the fish cutting houses, patronize only a few cutting houses, or trust the brokers at the wholesale market.

Filleting by Hand

Filleting fish by hand is an art. It is also very hard work. The fish are generally brought to the filletters and the finished fillets are moved along either by conveyor belts or by helpers. The fish frame/rack remaining after filleting is a potentially valuable by-product that sees very little use at this time. (See fish-flesh recovery and deboning equipment, Chapter 9.)

The filletters generally remove the two fillets from the fish, and are not responsible for the skinning procedure; with or without scales, the fish's skin is part of the "gurry," which is also removed. "Gurry" is the fish industry's generic term for all fish wastes. Except in the smallest plants, the fish's skin is removed by any one of a variety of skinning machines (see later).

Wages in these filleting plants may be based on either an hourly or an "hourly plus production" bonus; the latter is based on quantity rather than quality and is essentially the modern legal equivalent of "piece work." In the United States it is illegal to have a straight production-based pay. The hourly wages must at least exceed the legal "minimum wage." We should note that a champion filletter can do five 20-lb fish in 4.5 minutes.

Detailed studies of the filleting process at Memorial University in Newfoundland by Pesi Amaria in the 1970s have yielded practical suggestions for filletters. Amaria has found that the ideal height for the cutting table is waist high; it is worthwhile to adjust all the workers' heights by using an adjustable floor stand. Tilting the cutting board away from the cutter by about 8 degrees, which allows the worker to follow his or her natural arm motion, also improves the yield per man-hour.

Amaria has also studied the cutting process in great detail. When cutting the first side fillet, the fish is well supported by the flesh on the other side and the backbone is straight, making it relatively easy to cut the fillet cleanly. When cutting the second fillet, however, the bottom of the fish is not supported and thus the backbone is curved. The head of the fish remains attached during hand filleting of most species of fish. The yield obtained when making these second-cut fillets may be 1–2% less than for the first-cut fillets. To deal with the problem of the crooked back, Amaria has suggested the use of a notch in the cutting board so that the fish can be cut on the second side with the backbone flat. Unfortunately, the slight extra amount of time required for positioning the fish in the notch also has an economic cost, although not as great as the 1–2% loss of yield.

To better understand the economics of yield, let us explore a fictitious case that highlights the importance of a 1–2% loss of yield. A 4-lb cod (relatively small) costs 30 cents a pound when drawn (gutted) and has a labor input of 10 cents. A 40% yield (considered good) gives 1.6 lb of flesh with a cost of $1.30 ((4 lb initial × $0.30/lb) + $0.10 labor). This means that the cost per pound ($1.30/1.6 lb) is $0.81. A 2% yield loss gives 1.52 lb of fish—which still costs $1.30—forcing up the cost per pound ($1.30/1.52 lb) to $0.85. These few pennies really add up in a highly competitive business with large volumes of fish being processed each year. For example, if 10,000 lb of fillets are cut each week, that offers $400 per week in additional income. With the fictitious figures we are using, it also means that a 2% yield loss leads to over a 4% decrease in profits.

These calculations do not emphasize the effects of fish fillet size. The labor input for cutting fish does not increase in proportion to the size of the individual fish. Thus, within a certain range of fish, the cost to cut smaller fish is higher per pound of product than for larger fish. Again "small" practical changes can have "large" economic effects in the fish industry.

Filleting by Machine

If there is an adequate and regular supply of appropriately sized fish of usable species, then cutting fish by machine can be faster and more profitable than filleting by hand. Some observers believe that the final fish product is also of a higher, more uniform quality.

However, establishing a machine-based fish processing plant is an expensive and difficult undertaking; so is maintaining such a facility. The equipment is large and in many cases has been designed so that it can also be used at sea on larger ships. The initial capital investment must also insure that the site has adequate space as well as appropriate electrical capacity and, if necessary water hook-ups. Thereafter, of course, the equipment will require maintenance and clean-up attention. A trained mechanic must be available at all times in the more complex operations simply to maintain and repair the fish processing equipment. This is not true of the staff of cutters; the number of these employees is usually modified daily to reflect the workload.

Fish cutting equipment has various technical constraints regarding the size and type of fish that can be processed. Machines do not always match the yield obtained by an experienced manual cutter. Most recently, a great deal of effort has gone into designing a new generation of computer-controlled fish cutting equipment. Many of these machines are manufactured by the Baader Company of Germany, which claims that the yield is now equal to or greater than the manual yield. The cost per fillet is less. Such calculations take into account the fact that employees operating the machinery earn far less than experienced fish cutters.

Mechanical fish cutting is more efficient if the fish have been sorted by size beforehand. Sorting equipment currently exists but could be greatly enhanced by further technical development. Ideally, the fish would be sorted by species. One of the more expensive current methods of size sorting is to weigh each fish individually. This allows the original load of fish to be separated into separate lines for each weight class.

It is more feasible economically to sort the fish by size. Current sizing equipment uses rollers that move further apart as the fish work their way down the line. Arbitrary separations can be reasonably obtained as smaller fish drop through sooner and larger fish drop through later. Accuracy is not as great as with weighing-based sorters. The dropping and shaking is probably not good for the fish. In any case, mechanical processing is still the most practical method for filleting the smaller fish.

Ungutted Fish

With species that have not been gutted (e.g., redfish), the fillet cuts, whether by hand or by machine, are made such that they avoid the belly flap and gut area. The yield is slightly less but is made up by the savings in not having to gut the fish. The belly flap area is generally not included as part of the fillets from gutted fish either. Redfish are not gutted at sea because their sharp fins makes them dangerous to handle during gutting. They are also a relatively small fish and are caught in large numbers at one time.

"J" and "V" Cuts

An additional cut is made on many species, such as cod. It is done coming in from the neck area to remove a line of "pin" bones running towards the tail. Depending on how the cut is made, it is referred to as a "J" or "V" cut. This provides an excellent source of pure white flesh meat for the deboner, and is the source of much of the minced fish currently available.

POST-FILLETING HANDLING

Skinning

Various machines exist for skinning the fish. Most skinners work by catching the beginning of the fillet and then cutting between the skin and the flesh using a vibrating knife blade between the skin and the fillet. Baader is currently working on a system to permit the fillets to be machine-aligned skin side down to expedite the process.

An unusual approach for skinning is that of the Trio skinner, which first freezes the skin surface and then cuts the fillet. This technique permits the processor to better control the depth of the cut being made and is helpful in cases where a process known as deep-skinning is desired. Deep-skinning involves cutting a few millimeters into the flesh, rather than directly under the skin, thereby eliminating some of the dark muscle along the lateral lines and subcutaneous lipid. In most fish, the dark muscle of the lateral line is fattier, and more likely to be contaminated with undesirable organic compounds; it also generally spoils faster. Orange Roughy, a New Zealand fish, contains a subcutaneous fat layer that must be removed because it is high in wax esters which cause diarrhea in humans.

Chemical Treatments

Following filleting (and if required, skinning), the fish may be treated chemically with a dip or spray of a polyphosphate and/or brine solution.

The brine may be from 1 to 9% salt. The brine treatment is supposed to firm up the fish and eliminate some of the "ragged" appearance following cutting, particularly hand cutting. The dip tank may actually represent an ideal source of bacterial inoculation of the now almost-finished product. Good manufacturing practice suggests that these tanks be emptied every 4 hours or so and that they be kept cool, particularly in summer. It is a procedure that should probably be more highly monitored in many plants.

The polyphosphate can be added in a tumble tank. The tumbler helps "beat" in the polyphosphate. The tanks are designed for the fillets to remain in the polyphosphate solution for 30 to 60 seconds. Abuse of this system is common, especially by leaving product in the tank during the lunch period.

Candling

Certain species of gadoids such as cod, pollock, and haddock are candled after filleting and dipping. A candling table usually consists of a translucent piece of glass (or plastic) with a strong light under it (usually white). The light shines through the fillet, making visible any undesirable marine parasites (see Chapter 14). The parasites can then be cut out by hand while the fillet is on the candling table. A fillet that is heavily infested with parasites is considered to be of very poor quality and may be discarded. If the parasites are particularly small, as they often are in the case of pollock, the work is difficult and tedious; it requires a great deal of patience. Inventors are continually trying to come up with a better system that would eliminate the eye fatigue associated with current candling procedures. Much of the work on products destined for Europe and North America is done in the Asian countries, particularly Korea, almost always by women. These women spend their entire day looking for and removing these worms, often using an eye loupe to help them find the parasites.

It is extremely important that fish with worms do not reach the consumer. The Icelandic fishing industry estimates that it costs them over $10,000,000 per year to ensure the absence of worms in their products. Therefore, researchers are actively seeking improved methods of detecting worms in fish. Hafsteinsson and Rizvi (1987) at Cornell have had some success using scanning laser acoustical microscopy. Other research groups have explored X-ray and infrared (IR) based techniques. No machine has reached commercial production; however, further equipment and processing development are anticipated.

The question of bones in fillets is one that is confusing to many consumers. The term "fillet" simply indicates the type of cut being made. The consumer can expect the absence of the main backbone, but the definition does not guarantee the absence of other smaller bones. The fillet must be labeled as "boneless" in order to imply an absence of bone. Even then, there is a bone tolerance in any official standard. Care must always be exercised when eating fish.

Operationally, finding bones is generally easier than finding parasites because of the greater contrast with the flesh. X-ray machines like those used in airports to examine luggage seem capable of detecting the bones but they are not able to detect parasites. Such machines are being used in some plants. A machine developed by Lumitech of Denmark uses the fluorescence of the bone for detection.

PACKING

The fillets can now be packed for shipping or freezing. In-line electronic weighing of samples is now more feasible and allows for better grading of odd-shaped materials such as fish. Design Systems of Washington State has also developed a machine that can create portion-control units using a water-jet knife under computer control. The same company has received a grant for developing an optical parasite detection system. Bulk packs are put up in various metal or plastic containers, or in plastic bags. Skin-on fish fillets should always be packed skin to skin or meat to meat to avoid the situation where the pigment of one fish might discolor the flesh of another.

The perishability of fish makes temperature control very important in packaging. The bulk (mass) of the fish and the initial fish temperature determine how fast the fish can be brought down to an appropriate temperature. The cooler the fish are kept throughout processing, the better, but the temperature must be short of freezing the fish. This can be accomplished both by rapid processing and by keeping the plant cool. The plant temperature must, however, be acceptable to the workers! It is important to chill (ice) the fish well. In many cases some additional salt and/or polyphosphate solution may be added to the fish at this point, but visible drip in these packages must be kept to a minimum. Besides being unsightly, this drip represents a yield loss to the buyer.

The plastic packs may be covered either with a hard plastic cover or a clear plastic film that is heat-sealed into place. The latter are often referred to as "Gen-packs." The fishing industry continues to argue the merits of metal versus plastic containers, and each material seems to have its strong advocates. The higher thermal conductivity of the metal is cited against the convenience (and price) advantages of the plastics.

The fish industry has been trying to move its products faster by increasing the use of air freight. Special containers and handling procedures are essential in this undertaking. The alloys used in airplanes are easily corroded by escaping fish juices.

Finally, a question concerning evaluation: what should be looked for when visiting a fish processing plant? First, check what happens at lunch times and breaks: Is product put away or left out? Is there a break clean-up crew? If the clean-up crew are working, are they contaminating product? Other questions concerning timed procedures have been referred to in this chapter. These management problems affect the quality of the final product, but are sometimes not addressed by management in a consistent way; that is, the procedure may simply have evolved over time or be an aberration that is happening the day of one's visit.

For an illustrative scenario, imagine a small fishing community in which a husband and wife both work at the same fish plant. The husband is at one end of the processing line (e.g., receiving) and the wife is on the finished product packing line. If the lunchtime is staggered so that the product that has been started is cleared from the lines before each production subunit can take its break, the husband and wife cannot eat lunch together. If there is a specific clean-up, that is, having people stop and put things away, ten or fifteen minutes of production are lost from all the workers. If both procedures are skipped, there will be a somewhat lower quality product. The fishing industry is still an "old" industry in many communities of the world. The concurrent sociological and psychological ramifications of this fact cannot be ignored.

ISOELECTRIC FOCUSING

What's in a name? The economic value of fish is, unfortunately, predicated on the name of a species rather than on its true food value or culinary appeal. Many experts—and even more consumers—cannot tell the difference between a high-value and a low-value species. This is an open invitation to economic fraud by the fishing industry.

Regulatory agencies feel responsible for protecting the buyer from such deception. As a result, they have worked to develop methods of analyzing fish and fish products for species identification. The most successful efforts to date have centered on species identification of raw product; identification of the species of already-cooked seafood seems to be much more difficult.

Electrophoresis can be used to characterize different fish species because of its potential for revealing many different bands of proteins. These bands migrate differently due to small differences in charge and/or molecular size. Unfortunately, the different migrations are sometimes due to small changes that have occurred in these proteins during or because of processing.

Isoelectric focusing is one method of electrophoresis: the supernatant, that is, the soluble liquid or drip usually obtained from the fish, is separated by this technique such that each protein stops migrating at its nominal "isoelectric point," which is that point at which the effective net charge is zero so that the material will not migrate. Conditions for obtaining samples, preparing them, and running them on the equipment have been fairly well standardized. A set of authenticated standards are generally run simultaneously to insure that small methodological differences do not become a source of error. Unknowns are compared to the standards and a positive identification can often be made. All of these characteristics commend isoelectric focusing for certifying species identification.

A negative result found during isoelectric focusing may represent the presence of another fish species; however, it may just represent a different subpopulation of the same species, or even a different processing treatment used on one species that has led to differences in the migration of certain bands. Thus, before any legal action is taken, it is essential to identify what the sample really is. That is, it is usually not sufficient to prove that a sample of "cod" is not cod by showing that its electrophoretic pattern is different from that of the library sample of cod; it must also be shown that it is really something else, for example, "cusk."*

Another line of attack in species identification is to take advantage of the antibody–antigen reaction. Techniques such as monoclonal antibody production yield sufficient quantities of a given antibody for a given protein such that a standardized test for the antibody's presence can be run. Each antibody of interest must be carefully tested and screened for its range of activity with respect to fish species. This technique also offers hope that progress can be made in analyzing cooked products. These test kits may also be quicker and cheaper than previously available techniques.

REFERENCES

Faria, Susan M. 1984. *The Northeast Seafood Book*. Boston: Massachusetts Division of Marine Fisheries.

Hafsteinsson, H., and S. S. H. Rizvi. 1987. A review of the sealworm problem: biology, implications and solutions. *J. Food Protection* **50**(1): 70–84.

Welsh, F. W., and R. R. Zall. 1979. Fish scales: A coagulating aid for the recovery of food processing wastewater colloids. *Processed Biochemistry* August: 23–27.

---

* According to recent studies, the organoleptic characteristics of cusk are such that when "mapped" in two dimensions, it is located between market cod and scrod cod. That is, it resembles either scrod cod or market cod more than the two cods resemble each other. Is a price difference really justified? A case can be made that there should not be such a difference.

5

Shelf-Life Considerations
for Fresh Fish

Flesh foods represent over 25% of every supermarket dollar spent on food purchases. But they are very perishable. It is no surprise, then, that there is a great deal of interest in extending their shelf-life to enhance distribution and consumer use.

Because the North American consumer is prepared to pay a premium to have fresh product rather than frozen product, the cost of freezing these products cannot be recovered. Ironically, good commercial freezing and distribution could sometimes yield a better product than is now available fresh, especially with poultry and fish. How many consumers purchase fresh chicken or fish and then freeze it in the home freezer which was never designed for this purpose?

In discussing shelf-life, we define the end of the satisfactory shelf-life period as the point at which the consumer is no longer willing to repurchase the product, that is, the consumer refuses to buy the product again. If we were forced to translate this into statistical terms, we would probably use the point at which 5% of customers reject the product for repurchase; but this cut-off is arbitrary. The important point is that the end of shelf-life is not a magical microbial number or TBA (fat rancidity) number; neither is it a specific color. Such research tools are often misused and misleading.

Two areas of background information are important. The first concerns the critical role of temperature control in atmospheric storage: keeping the temperature low affects both the microbiological and biochemical aspects of the changes in quality. McMeekin and Thomas (1978) have proposed that the general equation for spoilage is

$$\text{Spoilage rate} = b(T - T_0)^2 \qquad (5.1)$$

where b is a constant for the particular spoilage process, T is the actual temperature, and T_0 is the reference temperature determined experimentally. The significant point is that T_0 is *not* 0°C (32°F) but is actually -10°C (14°F). With this information, we can construct a relative spoilage rate table (Table 5.1).

Table 5.1 Temperature Dependence of Spoilage

Temperature (°C)	Spoilage Rate (Relative to −10°C)
−10	0
−2	64
0	100
+2	144
+4	196
+6	256

According to this table, even if we can prevent microbial growth by various treatments other than temperature, the biochemical changes are most readily influenced by the choice of storage temperatures. For example, Hermansen (1983) suggests that the amount of purge (drip loss) in red meats is higher at 5–7°C (41–45°F), at which temperature there is greater spoilage since the purge is an excellent medium for bacterial growth. Therefore, the expected shelf-life of a product either with or without modified atmosphere or chemical treatments may really depend on the product's temperature history.

The second area of background information is the nature of current research and the literature that it has engendered. A lot of the data and subsequent claims on shelf-life extension depend as much on the method used to measure the results as on the temperature and related conditions of storage. The choice of the experimental "control" is also critical to the results. For example, we recall that normal, good handling practice for fish includes keeping the fish on ice (0–1°C; 32–34°F) in a cold room (2–4°C; 35–40°F). If the experimental pack is kept in the same cooler, but is in a gas-impermeable bag at a higher temperature than the control (e.g., 4°C vs 1°C), the extension of shelf-life gained by the treatment may easily be offset by the loss of shelf-life due to temperature differences.

THE CHEMISTRY OF SHELF-LIFE

It is no secret that fish is highly perishable, that is, it spoils easily; this creates a greater challenge for optimizing fish use. Discussions of fish handling have highlighted the problem of perishability and availability of fish in the marketplace. These issues—and our perceptions of fish quality—are directly affected by the chemistry taking place during the fish's "shelf-life."

The standard handling procedure for finfish is defined as icing the fish at sea and keeping it properly iced throughout distribution and handling. Most research projects use this standard as the "control" when measuring shelf-life during other handling methods; in some cases, what is actually done in the industry is used as the experimental "control"—even if it is an inappropriate or undesirable procedure.

One definition of "shelf-life" is the length of time from the day of catch that fresh fish can safely be in the marketplace, that is, that the fish is not spoiled. This is a public health definition; we will redefine shelf-life from the consumer's point of view later. For the commercially major species from temperate waters, the shelf-life is generally between 14 and 17 days when the fish is handled with the aforementioned standard procedure. Under the same conditions, fish from tropical, warmer, water will last from 21 to 24 days.

Why is there this difference between the two types of fish? The temperate water fish live in colder water. They often have psychrotrophic and/or psychrophilic bacteria as part of their natural flora; the lag time for the growth of these bacteria may be shorter because of some pre-adaptation to colder conditions. On the other hand, tropical fish do not normally have these bacteria; the few that might exist have certainly not been growing. Most of the contamination of tropical fish comes from the handling system, that is, the ice, the boxes, and the boat. The greater change in temperature, from the water to the icing procedure, has a greater effect on the metabolism of the fish by slowing the post-mortem biochemical changes. The rate of a fish's enzyme activities is optimized for its own normal ambient temperature; a marked decrease in temperature also slows this rate significantly. In the absence of ice, tropical fish spoil much faster than temperate water fish because of the more rapid growth of non-psychrotrophic spoilage bacteria.

It is often difficult to duplicate field conditions in laboratory experiments because postmortem bacterial contamination of fish generally comes from extraneous sources. Scientific experiments in which the process is slowed down to examine fish spoilage represent a "best case" that may not be obtainable in the field. Many laboratory techniques are also not transferable to field conditions. It is impossible, therefore, to simply extrapolate any laboratory shelf-life results to field conditions.

TMAO.   The major bacterial species causing the ultimate spoilage of normally iced fish are probably the Pseudomonads and *Altermonas putrefaciens*. The latter used to be known as *Pseudomonas putrefaciens*, but the name was recently changed to fit more logically into the traditional classification scheme. These spoilage organisms can use the nonprotein nitrogen present in the fish, particularly compounds such as trimethylamine oxide (TMAO), to produce various volatile, aromatic compounds. Specifically, these bacteria produce trimethylamine (TMA), a volatile and odoriferous compound, from TMAO. The TMA is believed to react with fish fats to produce the typical spoilage odor associated with fish beyond their prime. White fish muscle is

Table 5.2 Spoilage Compounds of Fish

Compound	Odor	Probable Source
H_2S (hydrogen sulfide)		Cysteine
$(CH_3)_2S$	Cabbage	Methionine
CH_3SH	Musty	Methionine
Acetic, butyric, propionic and hexanoic acid esters	Fruity	Glycine
Serine, TMA, DMA	Fishy, ammoniacal	Leucine, TMAO
NH_3 (ammonia)		Amino acids, urea

Courtesy of Martin et al. (1982).

generally lower in TMAO, but usually produces more TMA than does dark muscle. Table 5.2 lists the breakdown compounds derived from other compounds in fish.

TMAO is present in most marine organisms. It also seems to be present in a few freshwater fish such as burbot (*Lota lota*), the only known freshwater gadoid.

Gadoids constitute an important class of commercial fish including the cods, haddock, the whitings, the hakes, the pollocks, and cusk. These fish and many other species seem to obtain the TMAO from their own diets. The amount ultimately found in a particular fish varies with its biochemistry. TMAO seems to be higher in fish from colder waters and from fish that are more active; larger fish tend to have more TMAO than smaller fish. We estimate that elasmobranch fish like shark may have 200–250 mg TMAO/ 100 g of flesh while gadoids have 50–100 mg TMAO/100 g of flesh. Red hake generally has the highest TMAO level among the gadoids.

The concentration of these compounds in fish flesh is expressed in a number of different ways and the conversions between the different forms of reporting the data are (Martin et al. 1982):

1 mg TMAO-N = 5.38 mg TMAO = 72 µmoles trimethylamine oxide (TMAO)

1 mg TMA-N = 4.22 mg TMA = 72 µmoles trimethylamine (TMA)

1 mg DMA-N = 3.22 mg DMA = 72 µmoles dimethylamine (DMA)

2.16 mg FA = 72 µmoles formaldehyde (FA)

TMAO, TMA, and DMA (dimethylamine) data are often expressed in terms of weight of nitrogen. This does not permit a direct comparison of quantities of material with other compounds such as formaldehyde (FA), which has no nitrogen. Therefore, we strongly urge that TMAO, TMA, and DMA results be reported in µmoles per unit weight (1 g, 10 g, or 100 g) of fish flesh so that the stoichiometry of the reactions can be more easily followed.

TMA is the normal bacterial breakdown product of TMAO. Levels above 15 mg TMA-N (1.08 mmoles TMA)/100 g of flesh are considered to be a sign of spoilage in fish. One of the routine ways to measure these breakdown products is to measure the total volatile base (TVB). Other compounds such as ammonia are included in this measurement, but the most rapid change in the value of TVB is attributed to bacterially produced TMA. The test involves volatilization of the bases; the volatile bases are then trapped in an acid medium and analyzed by titration with a base. Spoilage generally occurs at 30 mg N/100 g as measured by the TVB method, that is, the spoilage level with TVB is about twice the level for TMA alone. The fishy odor of TMA plus lipid is generally detectable when the TMA level reaches 4–6 mg N/100 ml, that is, before the actual spoilage point. A modified ammonia electrode has been used to measure TMA specifically, but results have not been sufficiently reproducible to allow the widespread use of this more rapid testing method.

The presence of ammonia in fish is quite natural. It comes and goes during the postmortem storage period and is thus not a good indicator of the quality of the fish. The presence of ammonia must be ignored as a shelf-life indicator in evaluations of cooked fish odor. Some evidence suggests that bacterial spoilage intensifies the sensory characteristics already associated with auto-lysis rather than actually creating new ones. Certain transient intermediate spoilage odors are observed in sterile fish that are not found in naturally spoiling systems. This may be due to inhibition or the masking of these compounds during spoilage; another possibility is that these compounds are being degraded by the bacteria.

The production of DMA and FA from TMAO is a separate biochemical reaction that takes place in some species of fishes, particularly the gadoids. It is much slower than the TMA reaction, but does occur during frozen storage at temperatures above $-29°C$ ($-20°F$). Frozen storage offers this reaction sufficient time as the TMA production is inhibited. The reacting FA yields an unsatisfactory texture that will be discussed in Chapter 7. DMA is much less volatile than TMA and thus does not contribute much to the off-odor of fresh fish.

Histamine.   Bacterially induced changes in fish lead to the creation of other potential spoilage compounds that may be of public health significance. For example, histamine may develop during the spoilage of some fish, particularly scromboid fish such as tuna and mackerel. Histamine is a breakdown product of the amino acid histidine, and fish normally have measurable amounts of this free amino acid. Histamine is a capillary dilator that can cause an allergic response in some humans. This reaction can be very extreme—even fatal—in some cases. The bacteria that produce histamine do not grow at proper cold-storage temperatures. The presence of histamine is a sign that the fish have been handled improperly.

Some of the bacteria responsible for the production of histamine are believed to be members of the Proteus family. Recent work suggests that other amine compounds might also be produced and involved in the toxic reaction. We cannot generally predict when fish spoilage conditions might lead to problems. This may reflect the fact that, in some cases, the bacteria seem to produce the most toxin under minimal growth conditions; that is, the more stressed the organism is, while still being viable, the greater the toxin production. Thus, marginal mishandling or mild abuse of fish may actually cause more histamine problems than does gross abuse.

For fish, a measure of biogenic amines, that is, histamine, putrescine, and cadaverine, may be a good measure of fish spoilage. In the United States, a level of 50 mg histamine/100 g for tuna and mahi-mahi, or dolphin fish, is considered a "hazard-action level." The general defect action limit for tuna and mahi-mahi in the United States is 10–20 mg/100 g. However, it should be noted that the presence of "potentiators" in the fish is required in order to cause histamine poisoning in humans at these levels. The level of these potentiators may be different under different circumstances and would presumably change the toxic dose of histamine required for symptoms to appear.

Ciguatera Poisoning.   Ciguatera poisoning is common to tropical fish. The source of this toxin problem seems to be organisms eaten by the fish, such as the dinoflagellate *Gambierdiscus toxicus*. Fish caught near reefs or in shallow water, particularly those with a good seaweed cover, seem to be more likely to be affected. Unfortunately, there are no quick methods for detecting potentially dangerous fish in the field. A great deal of effort is being directed to developing such methodologies to allow fuller exploitation of the available fish resources of the tropics.

Some ciguatera toxin may accumulate over time; the levels tend to be higher in bigger fish. The actual "ciguatoxin" is lipid-soluble but there may also be some water-soluble toxins that are found in fish that ingest dinoflagellates.

Proteolysis.   Proteolysis, or proteolytic degradation, another cause of fish spoilage, occurs with well-fed fish if the fish's numerous digestive enzymes come into contact with the fish's flesh. In fish like Pacific whiting, parasites containing proteolytic enzymes cause mushiness during postmortem storage. Since these enzymes are not activated until the fish are dead, these fish must be processed rapidly; the candling/parasite removal step must be done with particular care.

Work with fish minces, that is, a hamburger-like material that can be recovered from fish using a special deboning equipment, and surimi, a washed mince containing cryoprotectants (Chapter 9) suggests that some fish may have a heat-activated proteolytic enzyme with a temperature optimum of around 50–70°C (122–158°F). These enzymes seem to be similar to enzymes like papain and the other vegetable-based proteolytic enzymes. The consequences of this enzyme or these enzymes will become more significant as the industry moves into more complicated further processing of fish.

Nucleotide Breakdown.   The breakdown of the nucleotides of flesh foods is often used as an indicator of fresh fish quality. As with all flesh foods, the decrease in ATP (adenosine triphosphate) is part of the process leading to *rigor mortis*. First, ATP is broken down to ADP (adenosine diphosphate) and then resynthesized using stored high-energy phosphates, for example, creatine phosphate. When these are used up, the ADP itself is used to produce 1 mole each of ATP and AMP (adenosine monophosphate). Further breakdown of AMP leads to IMP (inosine monophosphate), and I (inosine), and eventually to HX (hypoxanthine). Subsequently, the HX may be metabolized further. Depending on the species of fish, the rate-limiting enzyme step in this breakdown sequence may be either the IMP-to-I step, in which case IMP will tend to accumulate, or the I-to-HX step, in which case I will tend to accumulate.

However, other aspects of this reaction are also of interest to the food technologist. The nucleotide breakdown pattern

$$ATP \rightarrow ADP \rightarrow AMP \rightarrow IMP \rightarrow I \rightarrow HX \rightarrow \cdots$$

affects the organoleptic qualities of a fish. The compounds that are flavor potentiators highlight (or are part of) the mild, slightly sweet flavor of fish. IMP may actually be sweet and may be an important part of the flavor associated with seafood. HX is a mildly bitter compound whose build-up can make a fish's taste unacceptable in the later stages of shelf-life. Various chemical methods of determining the different nucleotide breakdown compounds have been proposed, so that the change in these compounds might be used as a "freshness indicator." One test involves a rapid color reaction using an enzyme-impregnated piece of filter paper and an extract from the fish (Jahns et al. 1976). This technique is not yet available commercially. Improvements in chromatography, particularly high-pressure liquid chromatography (HPLC), have permitted faster and easier detection of all of the nucleotide breakdown compounds. Rather than relying on a single compound, researchers can determine various ratios of these different compounds with the hope of predicting fish quality more accurately. How are the results of these chemical tests evaluated organoleptically? To what should these results be compared? In other words, for any chemical test to be useful as a "fish quality" indicator, it will have to correlate with some type of sensory test (see Chapter 6).

As the reaction scheme indicates, HX is not the final breakdown

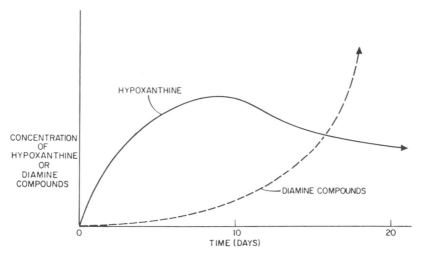

Figure 5.1 Time course of hypoxanthine (HX) and diamine development in fish. (*Martin et al. 1982, p. 355*)

compound. Its level will drop again over time, that is, there is a point in the postmortem storage period at which it peaks. Deciding which side of the peak one is on when measuring HX can be solved by measuring a second compound that does not appear until later in the spoilage process. G. Rand at the University of Rhode Island has proposed using the diamines, that is, putrescine and cadaverine, the "disagreeable" breakdown products, as the second indicator compound. They have already succeeded in developing a test strip assay for these compounds. Figure 5.1 illustrates the relative time courses for HX and diamines.

HX is produced at various rates in different fish: of those measured, it is fastest in redfish and slowest in swordfish. Two enzymes are involved: a hydrolase that may be autolytic and a phosphorylase that may be bacterial. Another test, the K ratio $(I + HX)/$total nucleotide (i.e., ATP, ADP, AMP, IMP, I, HX) or the K1 ratio $(I + HX)/(IMP + I + HX)$ may be more useful as a shelf-life indicator. Interestingly, species with a rapid IMP breakdown rate may well have a shorter acceptable period organoleptically. A number of instruments for measuring these ratios are now appearing on the market. One such instrument has been advertised that is supposed to measure K1. It has an electrode that is put into fish juice that has been treated with appropriate enzymes so that it measures the denominator of K1 and with a different set of appropriate enzymes it measures the numerator.

Lipid Oxidation.   Lipid oxidation is another important aspect of fresh fish spoilage and, indeed, all other types of fish storage techniques. It leads to off-flavors such as bitter and metallic, and to yellow-to-brownish discolorations. Flesh food lipid oxidation is generally measured by the TBA (thiobarbituric acid) test which uses a colorimetric procedure to measure malondialdehyde, a reactive intermediate, or by the PV (peroxide value) test which uses a titration procedure to measure the level of the relatively unstable early intermediary peroxides. As both of these tests measure intermediates in the oxidative rancidity process, both have severe limitations in the interpretation of test results. Further oxidation must be prevented during the testing procedures, possibly by the addition of EDTA (ethylenediamine tetraacetic acid), a chelating agent, or propylgallate, an antioxidant, before starting the TBA or PV tests. Oxidation may be slowed down by removing oxygen from the system or by using an antioxidant. Chelating agents are a specialized form of antioxidants that work by binding metals so that the metals are then unable to participate as catalysts in oxidation reactions. Since TBA values may reach a maximum and come back down again, the relationship of TBA to oxidative rancidity development becomes an "unknown" in a one-time measurement, that is, the test is much more useful in the laboratory to study known samples over time than in the field with an "unknown" sample.

Vacuum packaging or modified atmosphere storage, that is, storage in a specially created gas environment, for example, without oxygen, may be appropriate for extended storage. Interestingly, the presence of oxygen in the modified atmosphere is not detrimental to relatively low-fat fish. In fact, the presence of oxygen might even be desirable for the frozen storage of gadoid and other white-fleshed fish. Thus, the food technologist's normal reaction to exclude oxygen from foods whenever possible because of its potential to accelerate lipid oxidation may not always be the best procedure for fish.

In all cases, it is essential to keep track of the sample's temperature during handling and transport. New temperature recording devices are being developed for this purpose, such as the Control One, a waterproof box that can take up to 10,000 preset readings which are then read by computer. Others provide continuous stripchart records of the temperature.

Botulism. Botulism is caused by an organism that is found on and around many flesh food products. It is an obligate anaerobic spore-forming organism whose spores are difficult to kill. In the absence of competition, and possibly in specific microenvironments created by other bacteria, it can grow and produce an extremely lethal toxin. For example, even if the general environment contains oxygen, a local region directly below rapidly growing bacteria may become anaerobic, permitting the botulism organism to grow. Therefore, one goal of any shelf-life extension technique is that the product should spoil before toxin can develop.

The botulism organism can only grow at temperatures above 3.7°C (38°F). Thus, with proper handling it should be impossible for the organism to grow. Note that home refrigerators are almost always operated at temperatures higher than 3.7°C. The time for growth and toxin production at home refrigerator temperatures (less than 8°C; 46°F) is generally greater than ten days—less, of course, at higher temperatures. The spore levels, when they are present, in fresh fishery products are quite low.

Botulism testing procedures use much higher levels of spore inoculum: this inoculum is often injected into the flesh in a concentrated form. In fact, although the dose is calculated in terms of the entire flesh portion, the entire inoculum dose is injected into a single spot in the flesh. No shelf-life extension treatment can possibly reach the inside of the flesh where this inoculum has been injected. Test results are therefore doomed to give negative results. The use of a surface-swabbing application of the spores coupled with a "dose–effect" study is preferred. If the dose becomes too high, it "swamps" the ability of the protective compound or competition by other organisms to deal with *Clostridium botulinum*.

All botulism toxins are neutralized by the normal cooking of flesh foods; presumably this means that cooking until fish just turns opaque is acceptable. However, it is very dangerous to consume raw product if there is any chance that botulism toxin might be present. This fact contributes to consumers' concern about finding botulism in or on fresh flesh foods. However, with proper care to prevent recontamination, botulism toxin should not be a problem with fresh, unprocessed, cooked fish.

EXTENDING SHELF-LIFE

The extension of fresh fish shelf-life could have a significant effect on the fresh fish market in the United States and around the world. At the same time, food technologists could be improving fish quality in the marketplace.

A number of techniques have been explored for extending the shelf-life of fresh fish. In this context, "fresh fish" really only means fish that has not been frozen or otherwise processed; it is *not* a quality designation, but simply a market form. Consumers must understand that the word "fresh" does not tell us anything about how long the fish has been out of the water or its current quality other than that it is not "spoiled."

The results of studies on different shelf-life extension treatments depend on the quality of the initial fish material, the particular species chosen for the study, the market form examined, the temperature, and other handling conditions chosen to go along with the test variables. It would be wonderful to be able to predict the shelf-life of fish from one set of conditions to another. However, since current tests must be run for each species and set of conditions, it is impossible to use laboratory tests to predict what will actually happen in the "real world."

On the brighter side, laboratory tests can be good indicators of which methods show the most promise; with greater experience, we may be able to generalize more accurately. However, field testing must always be incorporated into a testing program for shelf-life extension techniques before a particular method can be recommended to the industry by researchers.

One question about fish shelf-life is whether fillets spoil faster than nobbed fish. The argument is that the cutting of the fish contaminates the fish flesh, making more fish juices available to the bacteria, thereby contributing to a faster spoilage rate for fillets.

Another shelf-life issue is the effect of temperature. This is a common concern for many biochemical reactions and microbial growth rates. Several researchers have reported that fish's response to temperature around the freezing point of water is essentially a monotonic function that can be fitted to a simple equation. If the temperature range of interest crosses certain temperature points, however, discontinuities can occur. Caution must always be used when extrapolating results from one temperature to another. For example, fish at 3°C (37°F) cannot provide conditions for *C. botulinum* growth; therefore, botulism toxin cannot exist at 3°C. One method of developing a set of temperature equations is discussed in Chapter 6. Other studies have focused on the ability of bacteria to function (and possibly grow? but not reproduce) at temperatures below 0°C (32°F).

Because different researchers conduct their studies at various temperatures, it is not always possible to compare data from different shelf-life experiments. At Cornell we use insulated ice chests in the refrigerator to minimize temperature fluctuations during experiments. This helps us interpret the experiments better, but unfortunately removes the results one step further from the real world. Note that researchers rarely monitor the fluctuations of their refrigerators; it is to be hoped that members of the fish industry do monitor theirs.

With these caveats in mind, we can now examine various shelf-life extension techniques.

CHEMICAL METHODS

A number of the techniques are chemical in nature; some of these also represent researchers' earliest attempts at extending the shelf-life of fish. Historically, antibiotic dips were used. However, these are now banned in most countries because of the fear that low levels of antibiotics in the diet may lead to changes in the microbial flora, eventually causing the growth of organisms that are antibiotic-resistant and detrimental to health. Various researchers claim to have proven that this already is the case with respect to antibiotics used in animal feeds. The results are quite controversial and have not been accepted by the poultry and red meat industries, particularly the latter, where antibiotics are still used routinely in feeds. The issue is of lesser concern in the fish industry since the use of antibiotics is generally banned. Only occasionally do samples with antibiotic residues show up in the commercial fish supply.

A number of other chemicals are being tested for shelf-life extension, including various compounds that are believed to be either bactericidal or bacteriostatic. The former actually kill the organism: the latter merely prevent their growth and may eventually be overcome. Standard chemical food preservatives such as potassium sorbate and sodium benzoate are used; in these cases, the active form of the compound seems to be the acid form. The pK (the pH at which the acidic and basic forms exist in equal amounts) for sorbates is about 4.7; for benzoate, it is about 1 pH unit less. Thus, as a system becomes more acidic, the amount of active compound becomes higher. However, if some of the active form is used up, for example, if it has participated in other reactions such as those that may cause it to kill or stop the growth of bacteria, the equilibrium will be re-established and a continuous supply of active form will become available according to the thermodynamic law of mass action. The concentration of the active compound changes with pH, but the amount of the potentially available active compound depends on the initial concentration applied. In the case of dips, this is the amount of material retained by the fish. Initially the concentration on the surface is higher, but with time, the material diffuses into and through the fish. With fish, of course, the final system pH is strongly affected by the total buffer capacity of the fish; lowering the pH, even at the surface, may also affect the texture and/or flavor of the fish.

The buffer capacity is a measure of how well a solution can maintain its pH when acid or alkali is added. Most flesh foods are reasonable buffers at physiological pH. The presence of small molecular compounds in fish, especially those with charge, adds to the buffer capacity.

One cannot assume that the effects of chemical changes such as lowering the pH are always perceived by consumers as negative. The subjective responses depend on the target consumers' reactions to the changes induced by the lower pH in texture and flavor. For example, some consumers may prefer the toughening of the fish texture that accompanies a pH decrease.

Dipping Procedures.   In any dipping procedure, the amount of chemical pick-up by the product is significantly lower than the dip concentration because the dip is diluted by the liquid of the fish. As a result, there is a dilution of the active agent; also, the pH of the solution approaches that of the fish itself. Eventually, even the solution in the dip tank comes close to the pH of the fish as it becomes contaminated with fish juice. This also means that the dip tank with fish juices may eventually be able to support bacterial growth as the active agent is diluted to below its effective concentration.

Dips are preferred by the industry because of their ease of use. However, single-use spraying would probably be far more sanitary. It would better serve the function of washing the fish while keeping the wash water clean.

Results of the dipping treatments may be affected by the postmortem pH changes of fish as it progresses towards its "ultimate pH" and then undergoes a pH rise as spoilage, for example, amine development occurs. The "ultimate pH" is the lowest pH reached by flesh foods postmortem. Metabolically the glycogen, animal starch, breaks down anaerobically to lactic acid. Different fish, that is, different species and individual fish within a species, reach different ultimate pH values. In general, because fish have less glycogen to start with and a lot of what remains is used up during the struggle prior to death, most fish have a higher ultimate pH than other flesh foods. As a rough guide, fish are generally above pH 6, while poultry is just below pH 6 and red meats will go as low as pH 5.2. The spoilage pattern and the effectiveness of the treatments varies depending on the state of the fish. For example, the higher the ultimate pH, the easier it is for bacteria to grow.

The timing of dipping is important. The maximum lowering of pH occurs while the fish are in contact with the solution; the highest concentration of the active ingredients must be picked up at this time. Once the fish are removed, the fish-plus-dip system is moving towards an equilibrium which dilutes the additive. The concentration of dip in the tank is also diluted by any liquid added to the tank by the fish being dipped. At the same time, we must assume that the bacteria on fish, like those on chicken, are found in "water pockets" along the surface of the product (McMeekin and Thomas 1978). Thus, the bacteria may well be protected from direct contact with the dip solution. There is a finite amount of time needed for the active compounds to diffuse through these water pockets and, of course, these compounds are further diluted in the process.

Fran-Kem is a commercial preservative compound available to the fishing industry; it is a mixture of sodium benzoate and fumaric acid. The former is a food preservative with a fairly low pK. The fumaric acid is presumably added as a food-grade acid to bring the pH of the solution down to the range where it increases the effectiveness of the benzoate. Two other dip products to extend shelf-life are Fish-Plus and FP-88. Fish-Plus is a mixture of polyphosphates, citric acid, and potassium sorbate. FP-88 is a mixture of polyphosphates, ascorbic acid, and salt.*

---

* With gadoids, the ascorbic acid might act as an accelerant of the gadoid texture change reaction during frozen storage since it serves to promote the TMAO to DMA and FA reaction.

The acidification process may affect other properties of the fish. Specifically, it may precipitate some of the proteins that ooze out at the cut surfaces, causing a change in the surface properties of the fish. Also, as we move closer to the apparent region of "minimum total charge" in flesh foods, estimated to be around pH 5.5, some of the water retention properties of fish are reduced. This loss may increase the drip: it is a weight (yield) loss and it is unsightly. The "minimum total charge" occurs at that point where a complex product has an equal number of positive and negative charges. This is sometimes mistakenly called the "isoelectric point." This term should be saved for individual proteins in specified solutions following an actual experimental determination of the pH at which a protein does not move in an isoelectric focusing gel electrophoresis experiment.

Shelf-life extension techniques may increase the drip loss simply by allowing more time for dripping to occur. Therefore, the very act of extending the shelf-life should be expected to increase drip.

Polyphosphates are included in a number of the newer systems for dipping fish. Sometimes acidic, they are added presumably to decrease the drip in addition to adjusting the pH. Polyphosphates probably affect water retention properties by their effect on pH and ionic strength, and by specific ion effects attributable to the various polyphosphate ions. Since these compounds are reasonably strong metal chelators, it is also possible that they have a small amount of antibacterial action and that they protect against lipid oxidation. Citric acid, another common additive, also chelates metal.

When selecting an acidulent dip, it is important to keep in mind its buffer capacity and its potentially toxic effect on the existing microbes. Its taste and taste interactions with fish are also important. And, of course, its odor must be acceptable to consumers. For these reasons, the use of a highly volatile and pungent-smelling acid such as acetic acid may be inappropriate.

Instead of dipping the fish into them, another way to use these chemical preservatives is to put them into the cooling ice. The slow melting of the ice yields a continuous application of the compound. However, since the distribution of the compound in the ice depends on the freezing process, the ice must be frozen quickly. Slow freezing concentrates the salt until the eutectic point of the system is reached, that is, the coldest temperature at which the solution can exist as a liquid, creating an outer layer of almost pure water with the inside of the ice having a much heavier salt concentration.

Even if the acid has a pleasant taste and smell, the final fish product may become unacceptable to consumers if there is too much of it. Too high a level of some of these compounds can lead to off-flavors. There must also be compliance with regulations that set a maximum of each chemical actually allowed in the food. If no specific tolerance is stated, then in the United States, the FDA "Good Manufacturing Practice" requires that the amount not exceed that which is "efficacious," or "needed to perform the desired function." If the additive is used early enough in the processing cycle, the FDA may permit the processor to consider the compound a "processing aid." Such aids do not require their use to be declared on the final label. However, the credibility of the fish industry is better maintained if a product's label includes information about the dip used.

IRRADIATION

Irradiation is the treatment of foodstuffs with ionizing radiation (gamma rays). In addition to extending the shelf-life of fresh fish, this technique seems to have other beneficial effects. Although high doses yield a very shelf-stable product—that is, one that does not spoil—biochemical changes can continue and cause unsightly light-induced browning. It is also likely that the gadoid texture changes that are normally associated with frozen storage might occur with long-term storage at ambient temperature (see Chapter 7). The latter processes, however, are far from receiving regulatory approval in the United States.

The operator of a fish plant using irradiation has many processes to control. In most cases the operation is done in a separate plant. Thus, attention to logistical details becomes a must. In addition to the irradiation procedure itself, proper packaging and temperature control must be maintained after treatment. Of course, the final product must also be consistent with consumers' expectations of quality. Otherwise, consumers would almost certainly blame irradiation rather than poor handling or marketing for a product's problems.

Although an FDA decision permitting the "general" use of low-dose irradiation is expected soon, the availability of the necessary facilities near the site where fish are landed and/or processed may present a problem. Commercial samples would have to be labeled to prevent fish from being re-irradiated. Whether consumer packages have to be labeled "irradiated" or not is an important question that awaits a final FDA decision. The initial FDA proposal published in the *Federal Register* was written so that labeling would not be required unless public response necessitated it. Numerous comments have already been received from the public—mostly negative toward irradiation or at least favoring labeling. The FDA has indicated that the labeling requirement will be kept at this time. We believe that irradiation will probably not be accepted at this time, in part because of the negative public response.

SUPERCHILLING

Researchers are working on superchilling or "partially frozen" storage of fish. At 0°C (32°F) water freezes, and until it is all frozen the temperature is unchanged. But, as will be discussed more fully in Chapter 7, a real food does not start to freeze until a few degrees lower. It also does not freeze at a single temperature, but rather more and more "ice" is formed as the temperature decreases. This intermediate range is referred to as "super-chilling," often involving specifically those temperatures just below the freezing point where no or so little freezing has occurred that no ice damage is found. At some point, the product legally becomes frozen. This system has been used quite extensively by the poultry industry.

General dogma states that freezing causes a texture change. However, it is not clear whether the problem is caused by the freezing itself or by the temperature fluctuations around the freezing point associated with mechanical refrigeration, which yields larger ice crystals and other texture-destructive changes. If it is the latter, then newer refrigeration systems might permit "crust freezing" to be used with fish. The poultry industry uses this type of system quite successfully. Even without freezing, one would presume that cooling fish to $-2°C$ (28°F), or just above the actual freezing temperature, should have a positive effect on extending biochemical life, even in comparison to storage in melting ice. This process of superchilling is receiving renewed interest and some supermarkets and distributors are re-examining this concept, particularly if the proper systems for maintaining superchill temperature with minimal fluctuation can be produced.

One of the properties that may be followed during storage at super-chilled temperatures is the amount of soluble protein extraction, a property that decreases with freezing—except that after 10 days, the amount of protein solubilized actually increases. For example, for sockeye salmon, 10 days at $-2°C$ (28°F) yielded acceptable quality; for sockeye and coho, holding at $0°C$ (32°F) gave acceptable quality. After that, the deteriora-tion in quality was rapid: in another 5 days, the fish were unacceptable or spoiled.

This spoilage pattern is characteristic of salmon. Salmon remain at an acceptable level for a long time, but then deteriorate very rapidly.

Superchilling is sometimes accomplished by using a blast freezer for a short period of time to cool the product quickly. The formation of small ice crystals with rapid freezing is an additional benefit of the quicker surface freezing. A new system, using very high-velocity air, is being proposed as a more efficient cooling system. We presume that the very rapid freezing of the skin of the fish will lead to minimal damage in the fillets. The system apparently would be capable of rapidly (in 10 minutes or so) bringing fish to a surface temperature such that the eventual equilibrium temperature would be 30°F ($-1°C$) without requiring any further heat removal.

MODIFIED ATMOSPHERES

How is a modified atmosphere created for seafood storage? The first concern is the gas used and the first choice is to simply remove all the air, that is, use vacuum packaging. This cost-effective alternative has been used very successfully in the boxed beef industry; beef primals (larger cuts of 20 to 40 lb) are now packed in plastic shrink-wrap vacuum packages and shipped that way rather than as hanging sides or quarters. After the vacuum is drawn, the remaining oxygen is used up by normal metabolism. If we start with a low microbial count, and keep the pH of the beef (approximately pH 5.5) low as well, then the product's shelf-life can be extended for a significant period of time. In fact, the biochemical changes taking place over time in beef can lead to improved product quality, that is, the aging of beef leads to flavor and possibly texture changes that are considered desirable by many consumers.

Vacuum packaging of beef also extends the time in which we can control the meat's color. During storage the meat's myoglobin (deoxygenated myoglobin) makes the beef purple-red throughout. However, upon exposure to air, it will rapidly regain its highly marketable cherry-red oxymyoglobin color. This color restoration process is often referred to as "bloom." Low levels of oxygen that are still higher than the levels of oxygen in boxed beef lead to the oxidation of myoglobin to metmyoglobin. This causes the undesirable and less reversible brownish color that must be avoided in red meats and poultry, and probably with the redder fish.

Nitrogen. Of course, creating modified atmospheres often involves adding certain gases. The addition of nitrogen, a totally inert gas, to flesh foods produces the same anaerobic atmosphere obtained with vacuum packaging. Again, depending on the bacteria present and their metabolism on the particular product, this effect may or may not be desirable. For example, in flesh foods, particularly fish and chicken, the absence of air leads to a preferential growth of those anaerobic bacteria that often produce the most obnoxious odors.

Anaerobic systems are also presumed to decrease the rate of lipid oxidation that results in rancidity.

Carbon Dioxide. Another gas involved in modified atmosphere systems is carbon dioxide. Carbon dioxide can dissolve in aqueous solution to form carbonic acid, a mild, weak acid. Carbon dioxide is apparently toxic to many bacteria. However, not all microorganisms are destroyed by carbon dioxide. On the contrary, the growth of some organisms is enhanced in elevated carbon dioxide atmospheres: their competitors have been wiped out and their own growth is thus favored. Fortunately, the "good growers" in carbon dioxide generally produce relatively acceptable "spoilage" compounds such as lactic acid from the lactic acid bacteria.

Because carbon dioxide is relatively soluble in (e.g., fish) flesh, the relative amount of carbon dioxide—as well as its total volume—may vary in the atmosphere. The adsorption may also cause some collapsing of the packaging. Packaging will be dealt with more fully later. Carbon dioxide may also directly affect the flesh food by acidification, leading to changes in water retention properties such that the drip is increased. The solubility of all gases in solution decreases upon heating during cooking, and this may lead to undesirable "bubble" formation on the surface of the product. At the same time, note that we do intentionally put lemon juice, another mild acid, on fish for its flavor-potentiating effect. Use of carbon dioxide levels of 40–60% for fish seems to maximize its antimicrobial effect and minimize the negative textural and/or color problems.

The interaction of carbon dioxide with myoglobin leads to metmyoglobin formation, which may yield gill browning in fish and discoloration in meat. In fish we have observed that high carbon dioxide levels (60–100%) make the eyes look too cloudy.

Carbon Monoxide.   Carbon monoxide has been proposed as a color fixative in flesh products, particularly red meats, where it forms carboxymyoglobin. The carboxymyoglobin also seems to help prevent fat oxidation. However, there are many complications in handling product with this gas and even more concerns about its safety in the diet. Currently it cannot be legally used in the United States. Sulfur dioxide has also been proposed for modified atmospheres, but it tends to be toxic, corrosive, expensive, and allergenic. A number of similar gases have therefore not been considered promising and little published work has been reported that uses them.

Oxygen.   Let us consider oxygen in a completely different way. One of the goals of much of the modified atmosphere work with seafood has been to exclude oxygen because of its role in supporting rancidity development and bacterial growth. However, this may not always be an appropriate approach. Large gassed transportation vans, which already exist for carbon dioxide modified atmosphere movement of beef carcasses, have carbon dioxide backflushed into the entire van and are designed for whole sides of hanging beef. Although these containers are tested for their "leak-proof" qualities, they do in fact have a finite leakage rate. Significantly, since the initial loading of the modified atmosphere cannot be done by creating a vacuum, there is always some oxygen present in the atmosphere. Let us consider a maximum possible level of 20% oxygen. The presence of oxygen may also prevent anaerobic spoilage and botulism.

The idea of intentionally incorporating oxygen in fish processing was considered to be ridiculous; the fish's high level of unsaturated fats alone has certainly discouraged other researchers. However, the Torry Research Laboratory of Aberdeen, Scotland, followed up on our idea (Fey and Regenstein, 1982) and now makes the following recommendations for modified atmosphere storage of fish:

For low-fat fish
- 40% carbon dioxide
- 30% oxygen
- 30% nitrogen

For high-fat fish
- 60% carbon dioxide
- 40% nitrogen

Apparently, fish do not have problems with fat oxidation if the fat is predominantly membrane fat (as it is in the low-fat fish). High-fat fish tend to be more pigmented; in this case, the oxygen seems to affect fat oxidation and also to cause variations in the pigments.

An intermediate type of modified atmosphere packaging system that has been proposed for fish, and possibly poultry, is one that uses a very gas-permeable film around the product. The master carton contains a large gas-impermeable film so that the entire master carton can then be gassed. During commercial shipping and handling, the product is protected by the modified atmosphere. When the master carton is opened prior to retail display, the product can "breathe" normally and the package's air is brought to normal atmospheric conditions while any "off-odors" are released. The technical problem with this system has been that the product package must be able to survive the pulling of the vacuum and subsequent reflushing. The film around the product balloons when vacuumed because gas cannot escape quickly enough.

Other Chemical Treatments.   Modified atmosphere treatments of flesh foods can be combined with other chemical treatments such as the use of phosphate dips to minimize drip. With antibacterial chemicals such as potassium sorbate (KS) there is often an additive, or even a synergistic, effect. When two compounds with similar beneficial effects are added together, the results of the addition can lead to three possible situations: (1) The sum of the joint action is greater than the numerical sum of the two; this is "synergism." (2) The sum of the joint action equals the numerical sum; this is an "additive" effect. (3) The sum is less than the numerical sum.

Another role of a pre-dip might be to get some antibacterial compound onto surfaces of the flesh that might eventually not be in contact with the atmosphere, for example, two fillets that cover each other up. Early research results suggest that potassium sorbate can be particularly successful in the shelf-life extension of fish when used in this manner. A possible concern when using polyphosphates with carbon dioxide-containing atmospheres is the effect of the final pH on the activity of both the carbon dioxide and the polyphosphates; this is particularly true for alkaline polyphosphates such as tripolyphosphate, which is often used with fish. Both the carbon dioxide's and potassium sorbate's antimicrobial activities are favored by more acid conditions; however, the drip prevention role of some of the polyphosphates probably comes from the shift of the pH to more alkaline conditions.

These chemical additives can be added to the ice and then be included within the gassed system. The melting of the ice continually adds the antibacterial chemical while the traditional washing action of ice is retained. This method is appropriate for van-loaded fish but not for small packages. In the small packages, an essentially 100%-humidity atmosphere is created naturally and maintains the moisture level, which prevents drying-out of the fish.

Summary Issues.   In dealing with modified atmosphere shelf-life extension systems, there are important summary questions to consider before adapting a system—especially because we are concerned with "real world conditions."

- What are the effects of the age of the fish product at the time of packaging? In general, the better the initial quality, the better the product.
- What effect does removal of the product from modified atmosphere packaging have on the subsequent shelf-life of the product? Post-treatment residual effects have been reported. That is, the product may not begin to spoil until a few days after removing the product from the atmosphere. Presumably this reflects in part the lag phase of bacterial growth.
- What happens when the product is temperature-abused?
- What is the ideal product-to-gas ratio? Does the ideal gas composition need change if the product-to-gas ratio changes? To optimize the benefits of the gas, enough must be present to be effective. Because carbon dioxide is soluble in the product, the actual ratio may affect the ideal gas composition.
- Do different parts of a product react differently? For example, does whole fish react differently from fillets of the same fish?
- How important is pH? The growth of microorganisms such as *Altermonas putrefaciens*, a major spoilage organism, appears to be pH sensitive.
- Is there any hope of using modified atmospheres with smoked fish? This is currently being done in the United Kingdom. Packages contain a large excess of gas and the distribution chain through the supermarket is carefully controlled (see later).

The United States government currently does not permit any modified atmosphere fish to receive United States Department of Commerce inspection (see Chapter 16). The closest it has come is to approve the use of a program developed by Cryovac, that is, Cryovac's system may receive grading services. This program uses an appropriate gas-permeable film to wrap the fish (a master pack concept) and is based on a tight $-2°C$ (28°F) temperature control program. The major benefit of this program may be its tight temperature control. Therefore, modified atmosphere packaging of flesh foods is not currently viable in the United States.

In the United Kingdom, however, modified atmosphere systems were introduced in the early 1980s for use with fish packaging. The system, introduced by Marks and Spencer, is now widely used by supermarkets there. Other fish are being cooled in "champagne" systems where the bubbling of the refrigerated seawater uses carbon dioxide instead of air to insure good circulation. The use of CO_2 dry ice as a coolant in the poultry industry is considered by some observers to be a form of modified atmosphere system.

Although there is a lot of potential and interest in this area, appropriate research involving both the industry and academia are needed. Careful consideration of consumer quality concerns would be necessary to make the laboratory results more applicable to the real world.

Theoretically, another potentially exciting area would be the combination of modified atmosphere packaging with low-dose irradiation treatments. However, since the latter is not at all commercially acceptable, the project might not get very far.

In the longer view, there may well be a future for modified atmosphere packaging and related systems where distance, time and/or consumer desires require it. A fringe benefit of such activities is the creation of an environment in which better handling of product becomes the norm (Wolfe 1980). Particularly in the fish industry, such an environment is really the more pressing need.

SOUS VIDE

Carefully prepared recipes, often center-of-the-plate dishes such as fish, can be put into a vacuum package while hot, pasteurized, and then stored in the refrigerator for a period of time. The product can then be reheated quickly either in the package (a boil-in-the-bag concept) or reheated in a microwave oven. With good temperature control and good sanitation, the technology is very workable in institutional settings. However, the product is not commercially sterile, that is, it has not received a heat treatment sufficient to destroy all potential pathogens, particularly botulism. Thus, although it could be dangerous if the product is abused, by attention to additional control factors such as pH and water activity, additional barriers to spoilage can be created. However, the use of "multiple partial barriers" rather than one insurmountable barrier is a new concept that will require time to become accepted. Currently the United States FDA will only permit "food processing plants" to prepare such products. Restaurants and supermarkets are prohibited from making this type of product in their instore kitchens.

OTHER TREATMENTS

Ozone.   Another treatment being proposed for fish shelf-life extension is the use of activated oxygen in the form of ozone. Ozone is generated *in situ* and then added to the water supply. It may help limit the bacteria on the fish. Ozone certainly sanitizes the water, particularly seawater, which is one source of bacterial contamination. With this technique, however, there is a concern that the presence of ozone may accelerate lipid oxidation.

Hypobaric Storage.   Hypobaric storage of fish involves the use of a rigid bulk container in which a large partial vacuum is drawn. The composition of the atmosphere can be controlled to some extent; humidity is also controlled. They are currently designed as part of a reefer truck (tractor-trailer size) system. This technique is quite expensive, involves the use of specially designed units, and has not yet demonstrated impressive results with fishery products.

Blanching.   Kelleher (1982) and Toledo-Flores (1988) in R. Zall's laboratory at Cornell have developed the blanching technique, a fish equivalent of pasteurization. A quick dip (2 seconds) in hot (90°C; 194°F) water is used before refrigeration of the fish. The technique seems to work best on flatfish where the heat treatment does not seem to affect the fish's appearance; some visual defects such as scale loss may be seen on round fish. The process may simply insure that the fish are properly washed and the wash water is not a source of additional contamination.

Other Ideas.   The aforementioned techniques may be combined with the hope of yielding either additive or synergistic effects. One strategy that has not been tested to date is to use a dip in either potable water or seawater containing a preservative at the time of catch or first handling and then to further treat the fish with modified atmosphere packaging at the time of processing. This would seem particularly appropriate for salmon or other fish that are often caught in remote locations, and then processed in plants which may be thousands of miles from the original point of catch.

Problems that might arise with respect to botulism affect all of these techniques. Another challenge of extending the shelf-life of fish is that biochemical changes that create the qualities in fish that consumers like should not be affected by most of these treatments, that is, those aimed solely at microbial quality. These techniques tend to extend the lower end of the shelf-life period where the microbial changes start to be more significant than the biological changes (Chapter 6). The only technique currently available to slow the biochemistry of the fish itself is to cool the product to a temperature just above its freezing point.

PACKAGING

Regardless of the fish's shelf-life, the consumer will use his or her own standards to evaluate the fish product when used in the home. An important factor in consumer acceptance is the nature of the packaging used.

Packaging fish in gas is a challenge. For instance, if we use a tight "shrink-type" pack we run into the aforementioned problems of anaerobic atmospheres. Fish packages with more gas have been more successful, particularly when carbon dioxide or other highly soluble gases are used. A gas-impermeable bag with the product on a styrofoam tray does not make for a particularly attractive package because of the resulting "balloon" effect.*

The containers and wrappers used in many supermarkets for individual ready-to-sell fish packages are often the traditional meat package with the styrofoam tray colored blue instead of white. Some of the form-and-draw type of rollstock films, particularly with a clear plastic bottom, might be particularly appropriate for fish. Although some packages are microwavable, as yet none is ovenable. Ideally, consumers will not have to handle the fish once they buy it; these packages would already be seasoned and/or sauced and could therefore compete with frozen entrees.

The use of diapers, that is, absorbent pads, with fish and other meats at retail is controversial. These pads pick up any drip that may form, but do they then encourage spoilage? Do consumers like them? Do they make the

---

* For air shipments it is best to under-gas the package. The lower pressure in the airplane, equivalent to about 8,000 feet even when pressurized, will cause the package to "balloon" during transport.

product look like something is being hidden? All of these questions require better answers than are currently available.

There is also the question of how this drip is taken into account when the product's net weight is determined. For practical reasons, both the federal government and the professional organization of people at the state level who ensure compliance with weights and measures laws have accepted that the drip should be included as part of the net weight of the product. This permits an inspector visiting a retail store to check for weight compliance by using a dry tare, that is, the packaging container with the diaper and the overwrap serving as the control, and then simply measuring the total weight of the unopened package to check for compliance. If one were to correct for the drip, the package would have to be opened, given time to drain, and then weighed. The latter is both time-consuming and destructive.

Any discussion of fish odor problems leads to the question of odor during cooking, considered objectional by numerous consumers. When fish is cooked, a lingering odor may permeate throughout the house. One possible advantage of the aforementioned ovenable tray and other new packaging is that it might minimize the amount of odor released into the home.

Ovenable paper and, more recently, plastic have been used in limited ways for various frozen seafood products. However, the recent development of upmarket frozen entrees suggests that consumers' attitude towards this type of packaging system may be quite positive. Could fresh fish be a prepared convenience product with the seasoning and/or sauce already included? Would this product benefit from gas packaging? How might temperature be maintained? That is, can they deliver the convenience of frozen food with the quality image of fresh fish, and really deliver a fresh fish that avoids problems such as handling, cooking, house odors, and bones? Culinova, a General Foods upmarket product, tried a similar approach with already-cooked fish. The product has been withdrawn, mainly because of the high price and the difficulties of maintaining the high-quality, labor-intensive distribution system that the product required.

Rigid packaging seems to be the best solution, for example, the "form and draw" type of trays, formed on equipment such as a Dixie-Vac, Swiss-Vac, or Multi-Vac. Many good gas-impermeable films do exist, and their slight leakage rate is not a major problem. On opening these bags one may detect a "bag odor." In general, if one slits open the bag and puts one's nose up to it quickly, one may detect an objectionable odor. However, if one pulls the product out of the package—certainly, the most usual procedure—the odor is not detectable. In many cases the odor in the bag is simply a concentrated version of acceptable, normal odors. Consumer perceptions demand that the industry continue to seek improvements in fresh fish packaging.

The ideal tray type of product would be made of a formable, gas-impermeable film that is both microwavable and "ovenable." The consumer would just make a few small holes in the film to prevent steam build-up or the film would release when heated sufficiently, that is, when enough heat and/or pressure builds up the seal between the top film and the main package separates. The consumer would put the fish directly into the oven or microwave. The bulk of the odors (spices and cooked fish, not raw fish) would be released at the end of the cooking time when the package is opened. This technology would also permit us to develop many different seasoned/sauced convenience fish products that could compete with other frozen prepared flesh food products. This type of system would address a number of consumer concerns: the lack of desire to handle the product, the problem of cooking odors in the home, the lack of knowledge of how to cook fish, the fear of overcooking fish (the moist heat minimizes this possibility), the need for convenience and portion control (Bisogni et al. 1987).

Containers and Packaging for Shipping Fish

Many different ways to package fish have been developed to accommodate the many different forms of fish that are shipped. The traditional wet-lock box (wax-coated cardboard) is often used for shipping fish on ice. Countries like Great Britain use wooden boxes for shipping certain types of fish. Some frozen-at-sea fish is not packaged at all but simply handled as is; sometimes a glaze, that is, a coating of water, is all that is used to protect the product. Dried fish in many areas of the world is handled in burlap bags.

In many ports the wooden, plastic, or metal boxes used to initially unload the fish are also used to ship and distribute the product. These containers may be designed to let any melted ice water drip out of the container or to insure that it stays inside. In the latter case, the boxes are often lined with a plastic bag when they are used for shipping. But fish have quite sharp fins and this can lead to leakage problems, even with plastic bags.

As the industry has moved towards more air shipping of fresh fish, leakage has become a serious problem. The drip from fish is highly corrosive to the aluminum and other high-tech metals used in airline bodies. Reports of seven-digit repair costs per airplane due to the leakage of fish juices have encouraged the airlines to become more selective about the types of containers they authorize for use.

New materials and designs are needed to improve container insulation. Fish must be maintained at an appropriate temperature throughout the distribution chain—no mean feat, considering the long waits between flights; at many airports there are no cold-storage facilities. As the airlines ship more fish—which, despite its problems, is quite lucrative—they will develop more airport refrigeration facilities.

Some of the larger shipping containers are designed as passive, for example, using carbon dioxide, cooling units while others use mechanical refrigeration units. However, only a limited amount of dry ice, that is, carbon dioxide ice, can be used in the hold of an airplane because its escape may kill any live animals that are also being shipped. One company has developed an insulated airline shipping container that is supposed to maintain the product's temperature for 4 days and not add additional weight as compared to an equivalent traditional shipping container. The unit is pre-cooled and contains a refrigerant solution that helps maintain the temperature. The container is also supposed to be sturdy enough that it can be used by the airline on the back-haul for luggage. Although these containers are designed so they can be washed inside, the hope is that the inner packaging of the fish would minimize the need for this extra step.

The customer normally pays for the weight shipped, including the container weight. This cost has to include any back-haul of a special container provided by the customer in lieu of the airline's standard containers. The back-haul rate must be negotiated with each airline; because they usually carry luggage on the back-haul, airlines can often offer a very favorable rate, sometimes gratis, if the container can handle luggage.

Other technologies are being explored. One packaging company is working on a hard plastic box with styrofoam lining which can be broken down into a few pieces that can be shipped back more efficiently. A corrugated cardboard company has made progress on a container of corrugated paperboard with a styrofoam lining.

Gel packs are often used to provide a source of cooling during shipping. The combination of water and a gelatin-type material creates a water–ice coolant package that does not drip, even when thawed. The coolants are often dyed blue so that any leakage would be obvious. One company has recently miniaturized the system by making each unit about 1 inch by $\frac{1}{2}$ inch by $\frac{1}{4}$ inch thick and selling them in sheets, similar to the "bubble" pack material used to protect fragile items during shipping. In fact, the material serves that purpose as well.

Trucking companies that ship fish have to maintain refrigerated warehouse facilities as well as special trucks for carrying fish. Unfortunately, these trucks cannot be easily used to haul other commodities back to the seaport communities; their higher rates reflect the absence of a viable back-haul. Some companies have been tempted to carry various nonfood items, including "garbage." This has led to very unfavorable publicity and the federal congress is attempting to write legislation preventing this practice. Temperature control remains a problem when product is left "out in the sun" at loading, transfer, and unloading points; the best refrigeration system cannot make up for abuses caused by human carelessness.

As more further-processed fish is shipped, it may become desirable to use a variation of the modified gas packaging system, that is, to gas the master carton and use a very permeable film on the individual subunits. The gas pack is therefore under professional management throughout the trip and the consumer receives a package in a normal atmosphere.

REFERENCES AND FURTHER READING

Aitken, A., I. M. Mackie, J. H. Merritt, and M. L. Windsor. 1982. *Fish Handling and Processing*. Edinburgh, Scotland: Her Majesty's Stationery Office.

Bisogni, C. A., G. A. Ryan and J. M. Regenstein. 1987. What is fish quality? Can we incorporate consumer perceptions? In *Seafood Quality Determination*, D. E. Kramer and J. Liston, eds., pp. 547–563. Amsterdam: Elsevier.

Fey, M. S., and J. M. Regenstein. 1982. Extending shelf-life of fresh red hake and salmon using CO_2–O_2 modified atmospheres and potassium sorbate ice at 1°C. *J. Food Sci.* **47**: 1048–1054.

Hermansen, P. 1983. Comparison of modified atmosphere versus vacuum packaging to extend the shelf life of retail fresh meat cuts. *Proc. Rec. Meat Conf.* **36**: 60–64.

Jahns, F. D., J. L. Howe, R. J. Coduri, and A. G. Rand Jr. 1976. A rapid visual test to assess fish freshness. *Food Technol.* **30**(7): 27–30.

Kelleher, S. D. 1982. A thermal treatment for the extension of fresh fish shelf-life. M.S. Thesis. Ithaca, NY: Cornell University.

Martin, Roy E., George J. Flick, Chieko E. Hebard, and Donn R. Ward. 1982. *Chemistry and Biochemistry of Marine Food Products*. Westport, CT: Avi.

McMeekin, T. A., and C. J. Thomas. 1978. Retention of bacteria on chicken after immersion in bacterial suspensions. *J. Appl. Bacteriol.* **45**: 383–387.

Regenstein, J. M. 1982. The shelf-life extension of haddock in carbon dioxide–oxygen atmospheres with and without potassium sorbate. *J. Food Qual.* **5**: 285–300.

Regenstein, J. M. 1983. What is fish quality? An essay for the fish industry. *Infofish* 6/83: 26–28.

Regenstein, J. M. 1984. Some notes on training a Torry style freshness panel. *Infofish* 6/83: 42–43.

Regenstein, J. M., and C. E. Regenstein. 1981. The shelf-life extension of fresh fish. In *Proceedings of the International Institute of Refrigeration*, pp. 357–364. Boston.

Regenstein, J. M., and C. E. Regenstein. 1981. *The Shelf-Life Extension of Haddock in Carbon Dioxide–Oxygen Atmospheres With and Without Potassium Sorbate*. Torry Document 1527. Aberdeen, Scotland: Torry Research Station.

Silliker, J. H., and S. K. Wolfe. 1980. Microbiological safety considerations in controlled-atmosphere storage of meats. *Food Technol.* **34**(3): 59–63.

Toledo-Flores, L. J. 1988. Extending storage life of fresh tropical fish in ice/salt mixtures alone and combined with antimicrobial treatments. Ph.D. Thesis. Ithaca, NY: Cornell University.

Wolfe, S. K. 1980. Use of CO- and CO_2-enriched atmospheres for meats, fish, and produce. *Food Technol.* **34**(3): 55–58, 63.

6

Assessing Fish Quality

The discussions of fish handling in Chapters 1 through 5 have included descriptions of objective measures of evaluation of fish. Throughout, we have referred to "fish quality"—an amorphous condition, presumed to be desirable, but too vaguely defined.

Why is it so hard to define fish quality? In part, because there are two camps divided: consumers resent, or at least resist, attempts to quantify what they view to be a qualitative issue; researchers are equally reluctant to accept nonobjective, marketplace standards of organoleptic evaluation. The United States government's "standards" for fish products usually deal with evaluation of workmanship and simply refer to "flavor and texture characteristic of the species." The Code of Federal Regulations does not offer further details.

Fish quality cannot possibly improve until food technologists use a common vocabulary for its measurement and description. Without expecting to satisfy these polarized audiences, then, the authors suggest the following definitions.

1.   As discussed earlier (Chapter 5), the end of a fish's shelf-life is the point at which the consumer would not repurchase the product. This means that chemical and/or microbial criteria must be correlated with consumer perceptions. In many cases, the commonly used laboratory tests may not meet this criterion, or have never been tested critically in this respect, and do not correlate with consumer desires. This type of quality definition is tied to consumer perceptions; the end point for shelf-life may vary depending on the location and/or ethnicity of the consumer audience. A distinction can be made between these measurements of fish quality and objective measurements that relate to public health and safety concerns. If a fish's bacterial count becomes high enough, the product can and should simply be declared spoiled.

2.   Gradations of quality levels are also based on consumer perceptions. Conditions yielding "premium quality," as opposed to "regular quality," for one particular species of fish may well differ from those of a different

species of fish. Some fish, like many meats, need more time to develop their prime flavor; others taste better closer to the time of catch.

Although the characteristics of a "premium" flavor have been well determined for some species, for example, the gadoids, in many cases there is still no agreement between the marketplace's definitions of premium and supporting laboratory data.

LABORATORY INDICES

The terms "quality" and "shelf-life extension" both reflect human perceptions. This line of thinking suggests the use of sensory panels: this is difficult and costly, but generally accepted as the "bottom line" in evaluating food products. We suggest that any other chemical, physical, or microbiological tests that are run must correlate to sensory evaluation. The question is how this can be done. Consumers' perceptions of quality must be quantified to validate any laboratory results from shelf-life extension experiments.

The Torry Freshness Scoring System.   To quantify organoleptic perceptions, we suggest the use of an objective sensory panel and a system like the Torry Freshness Scoring System developed at the Torry Research Station, Aberdeen, Scotland. The Torry Freshness Scale uses a trained panel to critically, objectively, and reproducibly evaluate the organoleptic qualities of fish, that is, the "state of the fish." The panels determine the postmortem age of the fish (Table 6.1); they do not actually judge the subjective quality of the product. Under ideal circumstances, the Torrymeter, an instrument designed by the Torry Research Station, and now available commercially, can

Table 6.1   Torry Freshness Scores (for Cod and Other Gadoids) and Their Relationship to Other Measurements

Days	Freshness Score	Torrymeter	TMA[a] (μmoles/100 g)
2	9	14	7
5	8	13	22
8	7	11	137
11	6	10	360
14	5	8	792
17	4	6	1,730
20	3	4	3,240

Courtesy of Torry Research Station, Aberdeen, Scotland—Crown copyright.
[a] TMA = trimethylamine.

accomplish the same goal. This instrument measures either one or sixteen fish and gives a "Torrymeter reading" that can be correlated with the value derived from the Torrymeter Freshness Score. It works well with whole fish and whole gutted fish, but does not work with fillets. It is designed to correlate with the properties being measured by the human panel, and does not work as well with other score sheets.

The first step is to train a human panel to respond to specific criteria that are thoroughly worked out for a particular fish, for example, a whole dressed (gutted) cod. The "raw" score sheet takes into account many of the factors that people have been traditionally trained to consider when evaluating fish quality. But it also includes in great detail a few special factors such as gill odor, which is, in fact, the best indicator of changes in raw cod and haddock over time. Each score represents about 3 days of normal aging of the product (i.e., fish stored in ice). Table 6.2 shows the Torry raw freshness scoresheet, while Table 6.3 shows the cooked freshness scoresheet.

Both fish flavor and odor vary with time. If we look at the equivalent score sheet for the cooked fish, we find other technical criteria that permit the evaluation of the fish by a trained panel. For cod and haddock, flavor is the most important criterion. For these fish the highest score, that is, for the freshest fish, does not have the best organoleptic description. It is score 8 and 9 cod and haddock that we predict are really premium fish while score 6 and 7 represent acceptable quality cod and haddock. Thus, like many flesh foods, these fish also improve with a slight aging.

The Torry Freshness scoresheet has been used to develop product specifications for British trade that are enforced by the judgment of one or two inspectors. These objective measures of organoleptic qualities, as described on these score sheets, should allow any group of trained panelists to get the same score for identical samples anywhere in the world. Careful linguistic work and proper training of panelists must be used to establish equivalent verbal descriptors around the world. It is essential to train the panel well, for example, by working with samples of known history, to keep the panelists motivated, and to check on their performance regularly by using blind control samples. Of course, score sheets for "raw" and "cooked" fish must be kept up to date, and correlated, in relation to any changes of treatment applied to the fish. Preliminary results suggest that the sequence of changes remains the same for a particular property, though the rate of change may be slowed down. In some cases, different properties will change at different rates. For example, the gill odor may not slow down as much as other characteristics being measured in the presence of a carbon dioxide/ oxygen modified atmosphere.

In relation to the effects of cooking fish in different ways on consumers' perceptions; the Torry testing system is an extremely critical test because the fish is cooked without any condiments. Consumers generally accept deep

Table 6.2 Torry Raw Freshness Scoresheet for Iced Cod and Haddock

Score	Eyes	Skin	Texture and Effect of Rigor Mortis	Flesh and Belly Flaps	Kidney and Blood	Gills		
						Appearance	Odor	Score
10	Bulging, convex lens, black pupil, crystal-clear cornea	Bright, well-differentiated colors, glossy, transparent slime	Flesh firm and elastic. Body pre-rigor or in rigor	Cut surface stained with blood. Bluish translucency around backbone. Fillet may have a rough appearance due to rigor mortis contraction	Bright red, blood flows readily	Glossy, bright red or pink, clear mucus	Initially very little odor increasing to sharp, iodine, starchy, metallic odors	10
9	Convex lens, black pupil with slight loss of initial clarity		Flesh firm and elastic. Muscle blocks apparent. In or just passing through rigor	White with bluish translucency, may be corrugated due to rigor mortis effect	Bright red, blood does not flow		changing to less sharp seaweedy, shellfish odors	9
8	Slight flattening of plane, loss of brilliance	Loss of brilliance of color	Firm, elastic to the touch	White flesh with some loss of bluish translucency. Slight yellowing of cut surfaces of belly flaps	Slight loss of brilliance of blood	Loss of gloss and brightness, slight loss of color	Freshly cut grass. Seaweedy and shellfish odor just detectable	8
7							Slight mousy, musty, milky or caprylic	7
6	Slightly sunken, slightly grey pupil, slight opalescence of cornea	Loss of differentiation and general fading of colors; overall greyness. Opaque and somewhat milky slime	Softening of the flesh, finger indentations retained, some grittiness near tail	Waxy appearance of the flesh, reddening around the kidney region. Cut surfaces of the belly flaps brown and discolored	Loss of brightness, some browning	Some discoloration of the gills and cloudiness of the mucus	Bready, malty, berry, yeasty	6
5			Softer flesh, definite grittiness				Lactic acid, sour milk or oily	5
4	Sunken, milky white pupil, opaque cornea	Further loss of skin color. Thick yellow knotted slime with bacterial discoloration.		Some opacity reddening along backbone and brown discoloration of the belly flaps	Brownish kidney blood	Slight bleaching and brown discoloration with some yellow bacterial mucus	Lower fatty acid odors (e.g. acetic or butyric acids), composted grass, "old boots," slightly sweet, fruity or chloroformlike	4
3		Wrinkling of skin on nose					Stale cabbage water, stale turnips, "sour sink," wet matches	3

Table 6.3 Torry Cooked Freshness Scoresheet for Iced Cod and Haddock

Score	Odor	Flavor	Texture, Mouth Feel and Appearance	Score
10	Initially weak odor of sweet, boiled milk, starchy followed by strengthening of these odors	Watery, metallic, starchy. Initially no sweetness but meaty flavors with slight sweetness may develop	Dry, crumbly with short, tough fibres	10
9	Shellfish, seaweed, boiled meat, raw green plant	Sweet, meaty, creamy, green plant, characteristic		9
8	Loss of odor, neutral odor	Sweet and characteristic flavors but reduced in intensity		8
7	Wood shavings, woodsap, vanillin	Neutral	Succulent, fibrous. Initially firm going softer with storage. Appearance originally white and opaque going yellowish and waxy on storage	7
6	Condensed milk, caramel, toffeelike	Insipid		6
5	Milk jug odors, boiled potato, "boiled clothes"	Slight sourness, trace of "off" flavors		5
4	Lactic acid, sour milk, "byrelike"	Slight bitterness, sour, "off" flavors		4
3	Lower fatty acids (e.g. acetic or butyric acids), composted grass, soapy, turnipy, tallowy	Strong bitter, rubber, slight sulfide		3

From Regenstein and Regenstein (1981).

fried fish of a lower quality than fish cooked by methods with a more subtle effect on flavor.

Rafagnataekni Electronics of Iceland has developed a machine that grades whitefish in the production line. It uses the principles of the Torrymeter and operates at a rate of 40 fish per minute. The technical report describing this new piece of equipment notes that Torrymeter readings for cod and haddock vary seasonally: reaching maxima in April/May and December/January and minima in January/February and August/September. The Torrymeter can be a valuable tool, but it must be used for species and conditions that are verifiable.

Branch and Vail (1985) have attempted to develop a small computer system to ask questions about fish quality. Panelists' answers range from two to four choices and are entered into the computer. Many different properties are monitored. Table 6.4 shows their scoresheet. The idea is that the number of defects increases with time, regardless of the particular fish's spoilage pattern. The hope is that the resultant curve of "defect points" versus time is relatively linear, reproducible, and therefore useful. Preliminary data suggest that this may be true. Figure 6.1 shows some of their preliminary data.

If the Torry scoresheet reflects consumer quality concerns, the second step of the process is to correlate the technical panel's results with the taste preferences of consumers as demonstrated in the marketplace. Work is needed to test whether the average consumer discriminates between a score 8/9 cod and a score 6/7 cod when offered the standard types of end uses,

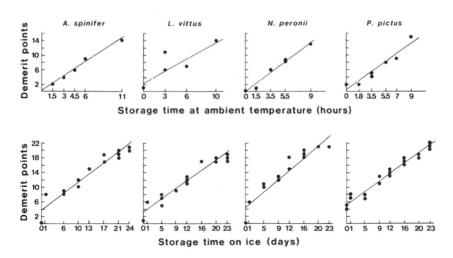

Figure 6.1 Preliminary data using the Tasmanian Defect Scoresheet. (*Kramer and Liston 1987, p. 430*)

Table 6.4 Tasmanian Defect Scoresheet

Fish Identification			Deteriorative Process
Appearance		(V. bright/Bright/Sl. dull/Dull) 0 1 2 3	Bacterial, chemical, physical
Skin		(Firm/Soft) 0 1	Enzymic
Scales		(Firm/Sl. loose/Loose) 0 1 2	Enzymic, bacterial
Slime		(Absent/Sl. slimy/Slimy/V. slimy) 0 1 2 3	Bacterial
Stiffness		(Pre-rigor/Rigor/Post-rigor) 0 1 2	Enzymic
Eyes	Clarity	(Clear/Sl. cloudy/Cloudy) 0 1 2	Physical
	Shape	(Normal/Sl. sunken/Sunken) 0 1 2	?
	Iris	(Visible/Not visible) 0 1	?
	Blood	(No blood/Sl. bloody/V. bloody) 0 1 2	Enzymic, physical
Gills	Color	(Characteristic) (Sl. dark) (V. dark) (Sl. faded)(V. faded) 0 1 2	Bacterial
	Mucus	(Absent/Moderate/Excessive) 0 1 2	Bacterial
	Smell	(Fresh oily) (Fishy/Stale/Spoilt) (Metallic, Seaweed) 0 1 2 3	Bacterial
Belly	Discoloration	(Absent/Detectable/Moderate/Excessive) 0 1 2 3	Enzymic, bacterial, physical
	Firmness	(Firm/Soft/Burst) 0 1 2	Enzymic, bacterial
Vent	Condition	(Normal)(Sl. break)(Excessive) (Exudes) (Opening) 0 1 2	Enzymic
	Smell	(Fresh/Neutral/Fishy/Spoilt) 0 1 2 3	Bacterial, enzymic
Belly cavity	Stains	(Opalescent/Greyish/Yellow-brown) 0 1 2	Chemical
	Blood	(Red/Dark red/Brown) 0 1 2	Physical, chemical

From Kramer and Liston (1987).

that is, various different ways of cooking cod and haddock. The concept of a premium fish for cod and haddock is theoretically clearly defined by the technical descriptors, specifically the distinctive, characteristic, sweet flavor. It is expected that this "clear" distinction will be affirmed. This simplicity does not always prevail. For example, the descriptors for cooked flounder are much more complex and, therefore, consumers' quality perceptions far more difficult to predict. The "premium" products may be of different postmortem ages, depending on the preferences of each local market. In such cases, consumer preference testing becomes the more critical second part of these studies. Tables 6.5 through 6.9 show additional Torry scoresheets for flounder, herring, trout, and salmon.

Scientific Measures of Fish Quality. A more scientific approach is to describe the changes that occur in fish over time by breaking the process down into various stages, for example, for gadoids (Figure 6.2).

Different species of fish have different time courses for the spoilage process, even under the same handling conditions. Concentration on a few significant species, therefore, helps to focus the research.

Table 6.5 Torry Scoresheet for Cooked Flounder

Score	Odor	Flavor
10	Meaty, oniony, fresh, salt butter or margarine, Worcester sauce, slight caramel	Meaty, very slightly bitter, shellfish, slightly garlic flavored
9	Oily, slightly aromatic, slightly peppery, boiled clothes	Oil, rather herring-like, metallic, but still meaty
8	Curry, still oily, peppery, damp clothes, baked smell	Curry, seasoned meat, oniony, spicy, peppery, canned meat
7	Caramel, boiled potatoes, musty, butterscotch	Neutral flavor, only slightly sweet and meaty
6	Metallic, slightly sour, acrid, slightly sweaty, boiled string	Slightly rancid, slightly sour, slightly bitter
5	Sour bread, lower fatty acids, rancid butter, singed milk, smoky	Rancid oil, rancid butter, fish meal
4	Slight amines, slight ammonia, sour beer, spoiled cheese	Bitter, woody, sour, little flavor
3	Ammonia, very sour, slightly fecal	Very bitter, rotten fruit

Courtesy of Torry Research Station, Aberdeen, Scotland—Crown copyright.

Table 6.6   Torry Scoresheet for Cooked Herring

Palatability

Odor

1. Fresh, seaweedy odor
2. Less fresh seaweedy odor, plus slight oily odor
3. Stronger, oil odor, often slightly "blown" plus slight "sweaty" odor, and sometimes "salt fishy" odor
4. Definite "blown oily," "salt fishy" and "sweaty" odors
5. Same as 4, plus stale, definitely unpleasant, ammoniacal, malty, sulfide or sour odors

Flavor of brown and white flesh

1. Fresh, sweet, seaweedy flavor
2. Less sweet, seaweedy flavor, plus slight oily flavor
3. Stronger oily flavor; some stale seaweedy flavor and some "blown oil" flavor
4. Definite unpleasant "blown oil," "sweaty" or rancid flavor, definitely stale
6. Repulsive flavor

Courtesy of Torry Research Station, Aberdeen, Scotland—Crown copyright.

Day 0	Acceptable quality
	↓
Day 2	Premium quality
	↓
Day 8	Acceptable quality
	↓
Day 14	Unacceptable quality
	↓

Figure 6.2   A simple quality time course for gadoids.

Table 6.7 Torry Scoresheet for Raw Trout

Days in Ice	Sensory Score	Eyes	Skin	Gills	
				Appearance	Odor
0–2	10	Convex, clear, bright	Clear, sticky slime; iridescent; silvery/green	Bright; clean; dark red to pink	Weak; trace of freshly cut grass, seaweed; slightly sharp
2–4	9	Convex to flat; very slightly gray; clear	Clear slime; slight loss of scales; iridescent; silvery/green	Bright red to pink; some bleaching	Some strengthening of the odor
4–6	8	Flat; slightly gray; clear	Slightly cloudy slime; slight loss of scales; loss of iridescence; silvery/gray		Shellfish; cut flower stems; fresh oil; aromatic
6–8	7	Slightly sunken; gray; slight opacity	Some loss of scales; slightly gritty	Pink to reddish/brown with bleaching	Cardboardy; aromatic malty; oily; acrid
8–10	6	Sunken; gray; slight opacity	Loss of scales; slightly gritty; dull colors	Pink to reddish/brown; bleached; slimy	Malty; beery; stale grass; slight rancid oil; putty
10–13	5			Pink to brown; bleached; sticky slime	Yeasty; stale beer; rancid oil; sour milk; boiled eggs; leather
13–15	4	Sunken; gray; opaque	Gritty; brown staining		Stale oil; lower fatty acids; stale vegetables; fruity

Courtesy of Torry Research Station, Aberdeen, Scotland—Crown copyright.

Table 6.8 Torry Scoresheet for Cooked Trout (Provisional scoresheet for flavor of cooked trout, steamed in casseroles without addition of water or condiment)

Score	Flavor
9	Metallic, sweet, fresh oil
7	Metallic, sweet, meaty, baked, slight herringlike
6	Meaty, chickenlike, cardboard, slightly sour, stale oil
5	Stale oil, rancid, sour, acid, cardboard, yeasty, mealy, woody
4	Nutty, cheesy, lower fatty acids, turnipy

Courtesy of Torry Research Station, Aberdeen, Scotland—Crown copyright.

The United Nations Food and Agricultural Organization (FAO) guidelines for shelf-life as expressed in terms of a Torry Freshness Score cut-off and as a function of temperature are:

(Torry Score 5 to 5.5)	
Temperature (°C)	Shelf-life (days)
0	14
4.4	6
10	3

Note that in the United Kingdom a score between 5 and 5.5 is used as the cut-off for "acceptable" and a score of about 4.5 is used as the cut-off for the product's being "unacceptable."

DETERMINANTS AND PERCEPTION OF QUALITY

How, then, do consumers evaluate fish? What qualitative tools do they use? What are the factors that determine perceptions of quality in flesh foods? Three essential components account for quality.

Table 6.9 Torry Scoresheet for Cooked Salmon

Score	Odor	Flavor
9	Baked, meaty, oily	Strong meaty, sweet, oily, metallic
7	Earthy, musty	Loss of sweetness, meaty, slightly musty
5	Musty, sour	Sour, musty
3	Sour, stale fruit	Sour, bitter
1	Rancid, sweaty	Putrid, nauseating

Courtesy of Torry Research Station, Aberdeen, Scotland—Crown copyright.

1. The *initial biological condition* of the material may be highly variable for wild caught fish, moderately variable for feed-lot beef, and almost invariable for a flock of chickens. The amount of control that can be had over this factor varies with the flesh commodity itself. However, the amount of control determines how well suppliers can manipulate the product to meet the needs of the marketplace.

2. *Workmanship* covers the many factors that can be controlled. For example, when harvesting fish, one can choose from among many different methods and thereby affect quality. Workmanship quality is affected by the way fish are caught, transported, and slaughtered. It is reflected in the entire processing, handling, and distribution of the product. It includes the care with which a product is treated: Have the defects been removed? Has the fat been trimmed? Are the pieces uniformly cut? Has the temperature been well controlled? These are all very much in the control of the industry and are often the basis for consumers' perception and evaluation of the final product. At other times, however, they are more important to the trade than to the consumer. Table 6.10 shows an example of the NMFS defect chart as an example of a workmanship assessment table.

3. Postmortem biology is sometimes referred to as "post-harvest physiology." These are the natural changes that occur in flesh tissue as well as the changes that occur because of microbiology. The flesh food interacts with its environment, from the composition of the atmosphere to the presence of light, and these interactions must be monitored for resulting changes in the final product.

Consumer Perceptions of Fish Quality. Consumers are becoming more sophisticated in their overall concern for nutrition. As a result they are switching to fish. Because they may not be familiar with fish, they are eager to learn practical method of evaluating fish quality. For example, consumers are taught to judge fish quality by studying the fish's eyes. This is effective if the fish has been held on ice with good drainage—and if the fish was never handled by its eyes! Many fish handlers find that the eyes are a particularly good place to get a grip on a slippery fish. Unfortunately, the biochemistry of the process which causes eye changes is not always understood, and the fish's eyes may not be an accurate indicator of its quality. Fishes have a permanently shut transparent eyelid. The lens itself is a hard sphere whose transparency depends on a specific water/protein/mineral balance. With constant washing, for example, in melting ice, the lens softens, swells, and becomes cloudy. The rate depends on the flow rate of the melting ice. The lens also becomes cloudy on freezing. Therefore, the eyes do not necessarily change at the same rate as other factors.

The authors are concerned that current evaluations of fish quality resemble the disparate descriptions of the proverbial "blind men exploring

Table 6.10 An NMFS Defect Chart for Fish

Factors Scored	Aspects Determining Score	Deduct
	Frozen state	
1. Color	Small degree: Moderate yellowing	4
	Large degree: Excessive yellowing and/or rusting	16
2. Dehydration	Minor: Moderate dehydration for each 10% of surface area affected	3
	Major: Excessive dehydration for each 10% of surface area affected	6
3. Uniformity of size	Minor: Each deviation from declared size in length, width, or thickness $\frac{1}{8}$ to $\frac{1}{4}$ inch	3
	Major: Each deviation from declared size in length, width, or thickness over $\frac{1}{4}$ inch	6
4. Uniformity of weight	Minor: Any minus deviation from declared weight of more than 1 oz but not more than 4 oz	3
	Major: Any minus deviation from declared weight more than 4 oz	8
5. Angles	Edge angle—2 out of 3 readings deviating $\frac{3}{8}$ inch	2
	Corner angle—each angle deviating $\frac{3}{8}$ inch	
6. Improper fill	For each 1 oz unit cut from the block that would be adversely affected due to air spaces, ice spaces, depressions, ragged edges, damage, or imbedded packaging material	1
	Thawed state	
7. Blemishes	Each blemish in 5 lb of fish block	2
8. Bones	Each instance of bones in 5 lb of fish block	5
	Cooked state	
9. Texture	Small degree: Moderately tough, dry, rubbery, or mushy	5
	Large degree: Excessively tough, dry, rubbery, or mushy	15

Grade	Flavor and Odor	Maximum Number of Physical Defects Permitted		
		Minor	Major	Serious
A	Good	3	0	0
B	Reasonably good	5	1	0
C	Minimal acceptable	7	3	1

From Volume 50, Code of Federal Regulations.

an elephant," where each experimenter is convinced that his definition of a small portion of the entity truly represents the total being. The hope is that more information and discussion will help lead to more agreement on definitions and methods concerning quality. We also recommend that the reader be aware of his or her own prejudices and misconceptions while studying conflicting reports from different research and consumer groups.

REFERENCES

Branch, A. C., and A. M. A. Vail. 1985. Bringing fish inspection into the computer age. *Food Technol. Aust.* **25**: 66–73.

Kramer, Donald E., and John Liston. 1987. *Seafood Quality Determination.* Amsterdam: Elsevier.

Regenstein, J. M., and C. E. Regenstein. 1981. The shelf-life extension of fresh fish. In *Proceedings of the International Institute of Refrigeration*, pp. 357–364. Boston.

7

Freezing requires lowering the temperature to $-18°C$ ($0°F$) or lower and is a popular method of preserving fish. But whether the process is carried out at sea or onshore, the fish's market value is lowered. Why? Are consumer perceptions (values) technically correct? The quality of the fish after freezing can only be as good as its original quality, but if done correctly, freezing may provide consumers, wherever they may be located, with the finest quality fish.

Prefreezing processing is generally the same for all whole fresh fish through the gutting stage. The gutted fish, or even whole fish, are then "neatly" placed into a vertical plate freezer and frozen. To enhance contact with the freezer plates, and to create a protective glaze, water can be added to the mass of fish. Since the added water increases the amount of material that must be frozen, this procedure is generally only used with the higher-fat fish, that is, where rancidity might be a problem. It is also effective with smaller fish where the packing is better because greater fish volume per freezer volume requires less water. Following freezing, the blocks are removed from the plate freezer, usually by simply sliding them out. Even without the water, the entire mass usually stays together because the skin of the fishes was originally so wet that ice bridges have formed. A common block size is about 45 kg (100 lb). The blocks are then stored in the frozen fish hold, often without any packaging. They may or may not be palletized at that point. These blocks are then off-loaded and moved to the processing plants—unfortunately, sometimes, in open trucks.

Freezing is an energy-intensive process. Good air circulation with a blast freezer or good contact with the cold surface is essential to ensure quick, efficient cooling. On the other hand, defrosting depends on the thermal conductivity of the fish, which is more efficient (i.e., heat conduction is more rapid) with the flesh in the frozen phase than in the unfrozen. That is, if cold (or heat) is applied from the surface, freezing will occur faster than thawing since the heat transferred to a fish from the room, which is the source or

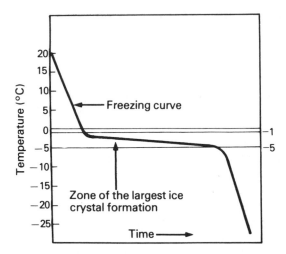

Figure 7.1   Freezing curve of products such as fish. (*Yamaha Motors, 1986, p. 44*)

sink of heat, is traveling through ice on freezing but through "water" on thawing.

The actual phase transition from water to ice requires much more energy than cooling. For example, to get from 5°C to −30°C (41°F to −22°F) takes 419 kJ (kilojoules) per kilogram of water, of which 333.5 kJ/kg is used for the phase transition, that is, the latent heat of fusion. As much as 321 kJ/kg is necessary to freeze a fish like cod (which is approximately 80% water). It is generally estimated that only 40% more energy is necessary to freeze tropical (warm) fish because of their higher ambient temperature.

The freezing of fish, unlike the freezing of water, does not take place at a single unique temperature, but is rather a gradual process taking place between −1°C and −5°C (30°F and 23°F) as shown in Figure 7.1.

Note on terminology.   We are using the vocabulary used by European researchers to distinguish between the functions of freezers and cold stores. A *freezer* removes heat from a product so that the product freezes. A *cold store* is a place designed to maintain the temperature of a storage area at a preset temperature but is not designed to do the extra work of freezing the product. Thus, a "home freezer" is more properly called a home cold store. Unfortunately, it is even poorly designed for its cold-store function. The set temperatures are relatively high, and there is too much exposed surface area, especially when the door is opened. The appealing self-defrosting feature also encourages moisture migration (see below) out of the product. (A chill store is a refrigerated storage unit.)

The ideal frozen storage temperature for fish is $-30°C$ ($-22°F$) or lower. Because of the lack of thermal equilibrium during the initial freezing, that is, because the inside of the fish is warmer than the outside, the recommendation is made to freeze the fish until the surface is down to -35 or $-40°C$ (-31 to $-40°F$) so that the whole block eventually equilibrates correctly without requiring any additional heat removal by the cold store. In the United States, most of the best cold stores are operated at $-20°C$. German factory trawlers apparently freeze fish to a core temperature of $-40°C$ ($-40°F$) and then store it at $-28°C$ ($-18°F$). Because the reaction of TMAO to DMA and FA will occur down to $-30°C$ ($-22°F$), the authors have been advocating storage below this temperature.

The degree of traffic into and out of the cold stores, the effectiveness of the insulation, the temperature and quantity of new product brought in, and the degree of care with which the defrosting cycle is managed all affect the temperatures and temperature fluctuations in the cold store and, most importantly, the amount of fluctuation to which the product is subjected.

Fluctuations in cold store temperature cause more damage to fish than a carefully maintained but too-high cold store temperature because the change in temperature causes a change in humidity. Table 7.1 shows the carrying capacity of air for moisture at various frozen storage temperatures. Air at a lower temperature has a lower carrying capacity for moisture. Excess moisture in the air condenses as ice, particularly near the coldest parts of the freezer, the fans, cooling ducts, etc. When the freezer warms up, the moisture that returns to the air often comes from the product. If the product undergoes enough cycles, the results will be "freezer-burn." These cycles also decrease the efficiency of the cooler as ice builds up and interferes with the movement of the colder air from the coils and fans to the product.

Most of the heat loss in a cold store is due to losses at the outer surfaces of the store. Therefore it is important to insulate the cold store correctly.

Table 7.1 Carrying Capacity for Moisture of Air at Freezing Temperatures

Air Temperature (°C)	Carrying Capacity for Moisture ($kg/kg \times 10^4$)
-20	6,373
-25	3,905
-30	2,346
-40	1,379

Needless to say, doors should be kept closed except when the store is being used. Heat curtains, for example, hanging strips of plastic at the cold store door that allow for movement of people and products but otherwise block most air movement, can help. The shape of the store should minimize surface area: a circle is ideal, but we usually have to settle for a cube in the real world. A single, larger store is thus more efficient than numerous smaller ones.

Processing Frozen Fish in the Factory.   *Thaw rigor* refers to the extremely strong contractions that occur during thawing of some flesh products, for example, frozen-at-sea fish, usually frozen pre-rigor. It causes a toughening of the flesh as well as significant water loss when the fish finally goes into *rigor mortis*. It occurs when flesh foods that have not had enough time before freezing, that is, frozen pre-rigor, for their ATP concentration levels to be lowered to the point where there is not enough ATP to support contraction. The high level of ATP remaining on thawing causes the muscle to contract. It is therefore best to thaw such fish slowly to a tempering temperature, usually about -2 to $-4°C$ (28 to 24°F). The fish are then held at this temperature long enough for nucleotide degradation to take place while the fish is still frozen. After these fish are fully thawed, they can be filleted and used for blocks, IQF (individually quick frozen) packs, or shatter-packing. The damage to the fish's texture and water retention properties due to freezing twice should be minimal.

Shatter packs are blocks of fish that are packed with subunits, usually fillets, wrapped individually in plastic film. The "masters" are usually 5 or 10 pounds. When dropped, the individual units break apart ("shatter") and can be sold separately. This form of packing requires extra handling, extra packaging materials, and the ability of workers to create reasonably equal-size portions that "make weight," that is, add up to the proper master weight.

Some processing plants use fresh fish for filleting and further processing into frozen products. If frozen product is eventually sold as fresh (thawed) product, it must be labeled "previously frozen."

Fish such as tuna that are destined for the Japanese sashimi (raw fish) market must be kept in excellent condition if they are to be acceptable and/or earn a top price. The fatter the fish, the better. Frozen storage temperatures of -55 to $-60°C$ (-67 to $-76°F$) are generally required in Japan for these fish.

A fish block is a carefully packed unit of fillets. The weight will usually range from 5 to 12 kg (11 to 27 lb). Minced fish is sometimes added to the block, depending on customs and legal constraints. Because the block is often subjected to mechanical cutting operations, it is important that the edges be straight and that there be no voids, that is, air pockets, in the product.

Figure 7.2   A Torry continuous belt freezer. (*Aitken et al., 1982, p. 64; Crown copyright*)

Efficient packing also helps to keep all of the blocks at their appropriate weight without waste. Putting up a good fish block, that is, one that shows good workmanship, requires constant management attention. These blocks are frozen in forms in a horizontal plate freezer. The blocks are then removed from the forms by a hydraulic ram after being fully frozen. The packaging for blocks, and many other fish products, is often a simple, inexpensive wax-coated carton. This is not an ideal package for the product because it fails to prevent air circulation through the product; the boxes are not tightly sealed and dehydration can occur too easily.

Each individual fillet or portion of IQF fish products is frozen in a nitrogen, carbon dioxide, or air freezing tunnel. This approach is more often used for further processed fish following batter/breading or similar operations.

"Belt" marks are sometimes left on the bottom of fish when a belt freezer is used for freezing the product. The impression of the belt is left because the fish's own weight is pressing down on the soft flesh that rests on the belt itself. The Torry blast freezer uses a continuous stainless steel sheet to carry the fish to avoid this problem (Figure 7.2). This freezer has also been specifically designed to give excellent heat transfer properties from both above and below the belt of fish so that the fish's thermal conductivity becomes the rate-limiting step. Because freezing takes time (the effect of

Table 7.2 Cold Storage Life of Frozen Fish

Fish	Storage Life
Fatty, >5% fat	
Herring	
Mackerel	
Salmon	3 months
Lake trout	
Lean, <5% fat	
Cod	
Haddock	
Flounder	6 months
Red Snapper	

From Ryan and Regenstein (1985).

thermal conductivity again) and space for large freezing equipment, space-efficient spiral freezers have become increasingly popular for commercial use. The Frigoscandia spiral freezer processes 4,000 kg (1,800 lb) of cod in an 8-hour shift with a dwell time of 20 minutes per fillet; the final temperature is $-35°C$ ($-31°F$).

With fishery products such as tuna and crabs, the raw materials may be brine-frozen, since brine has good heat-transfer properties. The samples are frozen and, in the case of tuna, possibly stored in this medium. Recent consumer concerns about salt levels in all food products has led to some reevaluation of this technology. One approach to reducing salt levels that shows some promise is to add calcium and magnesium salts to the brine solution. These cations seems to inhibit salt uptake. However, if added at too high a level, magnesium salts may lead to bitterness.

A proposal for the ultimate efficiency in freezing is to design special aircraft that would use the coldness of the upper atmosphere to freeze fish while on their way to market! This has been given some serious attention, although it currently seems rather farfetched.

Various charts exist giving recommendations for storage of fish. An example is shown in Table 7.2. It is important to identify the quality evaluation criteria, and consider whether they are relevant.

THAWING

The technology of the thawing of fish, that is, bringing the temperature of the product back above $0°C$ ($32°F$), is not as well developed as that of freezing. Many different thawing techniques are available and routinely used, each with distinctive advantages and disadvantages. Note that the rate of freezing is considered important because of its effect on ice crystal size (along with the temperature fluctuations during frozen storage), but there does not seem to be any evidence that the rate of thawing has any effect on fish quality, other than the effect of time.

The cheapest way to thaw fish is to put them into a running cold-water bath. Good water circulation is essential. The water can be run into pipes placed along the bottom of the thawing tanks with appropriately scattered holes to insure that many separate water streams are formed. Good mixing should take place as long as too much frozen product is not put in at one time. It is also a good idea to have an attendant break up the big fish blocks, for example, frozen-at-sea fish, as soon as sufficient thawing has occurred, to increase the number of heat-transfer surfaces. Because of the relatively long period of contact with water, this method is only recommended for whole, dressed or nobbed fish. The amount of water pick-up depends on both the time in the water and the water temperature. The amount of pick-up seems to increase as the temperature is lowered. Thus one report indicates a gain of 2.8% at 5°C (41°F) while the same sample lost 0.28% at 15°C (59°F).

The Dynajet is a new piece of equipment for thawing that uses a specially built cabinet with dripping water. The closed system offers better control, uses less water, and is cleaner than the more open versions of drip thawing used by the tuna industry.

Other plants use still air or moving air to thaw fish. Humidity control is important to prevent the product from drying out. The temperature of the surface of the fish must be kept low enough to minimize the growth of microorganisms.

The Torry Research Station of Aberdeen, Scotland, has developed a vacuum fish block thawer that works on the principle that the boiling point of a liquid is lowered in a vacuum. The boiling temperature of water in the vacuum drawn by this equipment is about 30°C (86°F). The heat to boil the water, that is, the latent heat of vaporization at that temperature, is the heat that can be removed from the frozen fish block.

The process of freezing can be reversed to create thawing conditions, especially when working with uniformly shaped fillet or mince blocks formed in a plate freezer. It is simply a matter of using a contact plate with running warm liquid instead of a refrigerant.

Dielectric thawing and electric resistance heating are two new techniques that have been developed in the laboratory pilot plant. The former uses a high voltage alternating current field, and the latter a low-voltage current through the material to be thawed. Although laboratory-based literature has emphasized the advantages of these methods, they have not yet been widely adopted in commercial practice.

Fish are often treated with a polyphosphate to help retain moisture when blocks are made. The process adds a little extra liquid since the polyphosphates are usually added as a solution. This addition of extra water is often the reason why regulatory agencies may oppose the use of polyphosphates— is the supplier just trying to sell more water? The polyphosphates also aid in binding the pieces of the block together either by solubilizing protein or by causing swelling of the myofibrils. Their antioxidant action during storage is probably due to their metal chelation properties, all of which affects the fish's flavor.

FROZEN FISH PRODUCTS

Fish blocks are the raw material for the fish portions, or the fish stick/fish finger market. Portions are cut into even rectangles or more unusual shapes such as squared-off triangles. The blocks are generally tempered before cutting, that is, warmed up to -1 to $-3°C$ (30 to 26°F), but not thawed.

A tempering room is usually used. This means that product must be held in a 5 to 10°C (41 to 50°F) cooler for a few days. The process is time-consuming, expensive, and inefficient because it is difficult to get all of the product to an even temperature. Another method of tempering is to use a microwave tunnel. Equipment exists to do this process continuously or in batch mode.

Tempering is an excellent use for microwave systems because runaway heating is less likely than in standard thawing. In the microwave, water is much easier to heat than ice. Thus, if a small amount of water is formed, it will often be preferentially heated to boiling while the nearby ice remains as ice. This is referred to as runaway heating.

The block being tempered needs a few minutes in the microwave and then about an hour to come to temperature equilibrium. The final product can have a very uniform and predictable temperature, which optimizes further processing. It is also a quick, efficient process. The major drawback is the high initial cost of the equipment and the ongoing expense of replacing the magnetron, the source for microwave energy, on a regular basis.

Regardless of the method used for tempering, the blocks are then cut with a bandsaw. This can lead to as much as 5% of the total weight of a block's being "sawdust" when sticks are being produced. Alternatively, the first two cuts can be made with a guillotine knife or a circular knife, which helps minimize sawdust, sometimes referred to as "kerf." The latter equipment is similar to the electric slicing wheel found in delicatessens. The circular knife, which costs about $50,000, requires the use of extremely evenly tempered product such as created by microwave tempering, the equipment for which may cost more than $250,000.

The cutting operation can also yield a lot of broken fish pieces, which are generally used to make products like fishcakes.

Following cutting, the product may be battered or battered and breaded. It may then be quick fried for $\frac{1}{2}$ to 1 minute, depending on the product size, to set the coating or it may be fully cooked, that is, deep fried. Sometimes the product is simply refrozen without any frying. The technology of batters and breadings is in a constant state of experimentation by the ingredient manufacturers. The larger, crisper Japanese bread crumb has enjoyed recent popularity, but it still causes problems because it stratifies and breaks when used in the standard breading equipment.

Batter and Breading.   Current research includes development of battering/breading systems for frozen products that can withstand institutional kitchen processes such as microwave reheating, infrared lamps (which are used to hold products warm in restaurants), steam tables (especially since moist heat is not compatible with crispness), and oven heating, because most institutional kitchens do not have deep fryers. Manufacturers of further processed fish products must constantly select, maintain, and upgrade the batter/breading system being used as technology improves. The development of "susceptor" technology in the microwave has permitted food companies to develop products that will stay crisp even when microwave-heated. Susceptors are special materials, often metallized films, that focus microwave energy onto the product so that temperatures hot enough for crisping can be attained.

The actual amount of batter and breading permitted by the United States Department of Commerce for various products is listed in Table 7.3. It is a difficult measurement to take, requiring the careful removal of the breading from the fish. Alternatively, the amount of fish present could be measured chemically. It is hoped that the measurement of the rare amino acid 3-methylhistidine might be a marker of fish content in products; if so, the specific amounts of 3-methylhistidine per 100 grams of fish would have to be determined for each species. This modified amino acid occurs almost uniquely in the myofibrillar proteins of most flesh foods.

Marketing and Quality Concerns.   Recent sales of deep-fried batter/breaded products have declined markedly in the United States because of consumers' perception and concern about value. Are they paying fish prices for bread? What about the amounts of oil used for frying and the calories in what is supposed to be a lean and healthful food? There has, therefore, been new interest in tempura, beer, and other light batters whose lighter breading keep down the total weight of carbohydrate and fat added to the fish.

Foreign markets such as the United Kingdom and New Zealand have sold frozen fish portions with sauces, often packaged in a boil-in-the-bag. The manufacturers' processing includes adding the hot sauce directly to the fish

Table 7.3 Standards for Breaded Products

	Percentage Fish by Weight	
Product	U.S. Grade A	Packed Under Federal Inspection
Fish portions		
Raw breaded fish portions	75	50
Precooked breaded fish portions	65	50
Precooked battered fish portions	—	40
Fish sticks		
Raw breaded fish sticks	72	50
Precooked breaded fish sticks	60	50
Precooked battered fish sticks	—	40
Shrimp		
Lightly breaded shrimp	65	65
Raw breaded shrimp	50	50
Precooked crispy/crunchy shrimp	—	50
Precooked battered shrimp	—	40

Courtesy of National Marine Fisheries Service.

portion and freezing the whole package very quickly. Popular flavors include dill sauce, cheese sauce, and parsley sauce.

It is probably safe to say that certain other countries are ahead of the United States regarding flavor variety and packaging innovations such as boil-in-bags. Boil-in-the-bag cooking does not add any odors to a household; long-term storage is also convenient. The area requiring improvement is the handling of the hot bag after cooking. One possible help is a slot for inserting a fork to lift the product from the boiling water and hold the bag during serving.

If the fish product is fried in a processing plant, personnel must be on the look-out for signs of oil quality deterioration since a rancid and/or burned oil can ruin the product. The oil is generally used up when darkening is observed. It is essential to filter the oil, especially to remove breading/batter that has been dislodged from the product. The addition of unsaturated oils from the fish fat increases the problem of oil breakdown. The tanks that hold the frying oil require regular cleaning with both alkali and acid. If the cleaning is not done correctly, foaming of the oil and "blowing-off" (breaking off) of the batter can occur during frying. Some companies use the color of the oil or a simple tasting of the product as an indicator. Others use a titration of the free fatty acids to detect unsatisfactory oil. Newer tests and additives to help maintain oil quality are being developed.

The distribution of frozen fishery products in North America leaves much to be desired. Fish are generally more sensitive to frozen storage conditions than other products and should be kept colder. However, it has been usual to have a variety of products stored together such that fish products are subjected to both too high a storage temperature and too much temperature fluctuation. European cold stores are often operated at $-30°C$ ($-22°F$) or below; North American counterparts tend to operate at -15 to $-20°C$ (5 to $-4°F$). The situation gets worse as the product proceeds through the food distribution system. The supermarkets' freezers and especially home freezers are often poorly set and/or misused; fish products should be kept there for a minimum of time. More effort is needed to inform the consumer and the food handling/food service industry of the limitations of normal cold storage for keeping frozen fish products.

THE CHEMISTRY OF FROZEN STORAGE

A number of physical measurements affect fish chemically in frozen storage. The *Handbook of Fishery Technology* (Novikov 1981; originally written in Russian, with the English version published in India) has many detailed tables of important characteristics of fish. Some of these are summarized below.

1. The onset and length of *rigor mortis* depends on the ambient temperature. At 1°C (34°F), it takes haddock 35 hours to attain *rigor mortis*, after which the haddock remains in *rigor* for 70–90 hours before *rigor* is resolved. At 15°C (59°C), it takes haddock 2 hours to reach *rigor*, where it remains for only 10 hours.
2. The freezing temperature of fish ranges from $-0.50°C$ (31°F) for common perch to -1.5 to $-2.0°C$ (29 to 28°F) for cod. Saltwater fish have slightly lower freezing temperatures. The heat needed for the phase transition of water is spread out over a large range of freezing temperatures for foods such as fish.
3. The percentage of water frozen as a function of temperature for cod is shown in Table 7.4.
4. The heat capacity (kcal/(kg °C)) for cod at various temperatures is shown in Table 7.5.
5. Phospholipase activity in cod decreases in the early frozen storage period, that is, 8 weeks at $-30°C$ ($-22°F$).

Chemical Problems During Frozen Storage

Dehydration. A number of chemical problems can arise during frozen storage of fish. We have already mentioned dehydration when the surface of

Table 7.4 Percentage of Frozen Water in Cod

Temperature (°C)	Percentage of Frozen Water
−1	8.0
−2	52.4
−3	66.5
−4	73.0
−5	79.2
−10	84.3
−20	89.0
−40	90.5

From Novikov (1981).

the frozen product is exposed to the circulating air. Unless the relative humidity of the air approaches 100%, the product gives up moisture to the air by sublimation, that is, the direct conversion of a solid into a gas; this process is accelerated if the product is warmer than the ambient air temperature. The warming of the air by the product increases the air's water-carrying capacity, and the extra necessary moisture is immediately supplied by the product.

Rancidity. Rancidity, or fat oxidation, can be a problem, especially with the fattier species of fish, but it can be prevented by protecting the fish from

Table 7.5 Heat Capacity of Cod

Temperature (°C)	Heat Capacity (kcal/(kg °C))
−40	0.44
−20	0.62
−10	0.95
−5	2.45
−4	3.61
−3	6.34
−2	15.58
−1	24.54
0	0.99
2	0.87
20	0.88

From Novikov (1981).

oxygen, most simply by glazing. Good packaging can help, as can the addition of water- or fat-soluble antioxidants. Fish seem to have some natural antioxidant(s), possibly compounds like vitamin E, which help minimize the extent of the problem. Hwang and Regenstein (1989) demonstrated that the best antioxidants for menhaden and mackerel were the water-soluble compounds ascorbic acid and erythorbic acid. This may reflect the presence in the natural product of a fat-soluble antioxidant with which the added compounds were acting synergistically.

It is more difficult to add a chemical to whole products, that is, fish or fillets, than to a minced product. Needle injection systems are used routinely in red meats such as corned beef and ham; they comprise many needles with numerous holes in each needle. Equipment to use this system with fish does exist; the benefits remain to be demonstrated.

Metals are often catalysts for oxidation processes. Addition of a chelator (e.g., EDTA) to bind these metal ions (particularly iron and copper) can also help prevent rancidity.

Textural Breakdown.   In some species, for example, the gadoids, the chemical TMAO breaks down to DMA and FA, and this can lead to a textural breakdown in the fish. A fish product that has undergone these changes holds the free water loosely like a sponge, thus losing all of its moisture on the first bite. With subsequent chews it appears to be very dry and "cottony"; the effect has also been described as "spongy." The reaction seems to be accelerated somewhat by cutting the surfaces, for example, when filleting, and more so by mincing the fish. Many factors may be involved in mincing and the complete explanation of why this accelerates textural deterioration is still under investigation. It appears that the reaction may be freeze-activated, that is, the process of freezing (via crystallization-induced changes) itself accelerates the reaction. The presence of metal ion contamination may also accelerate the reaction.

We advise caution when interpreting data based on the measurement of the accumulation of DMA. The reaction of importance to texture is assumed to be the reaction of the FA with the proteins of the fish. Reference is often made in the literature to "myosin cross-linking" but no direct evidence has been offered to prove that the observed changes in the structure of myosin, the major functional and structural protein of the muscle cell, are directly responsible for the changes in fish texture. On the other hand, it has been suggested that the total amount of DMA and FA produced in the fish tissue over time is not as critical to texture change as the production of FA and its subsequent reactions at certain sites within the muscle structure. The reaction of FA with protein takes place after this step. If the ability of FA to react with protein is modified for any of a number of reasons, for example, by an FA-scavenging additive, this fact is not reflected in the DMA production. In fact, a reaction according to the law of mass action might actually cause the DMA production to increase.

Lundstrom at the NMFS Gloucester laboratory suggests that the presence of oxygen might actually inhibit the reaction of TMAO to DMA and FA. One possibility is that the metal ions in the fish flesh may scavenge the oxygen and thus not be available to accelerate the TMAO reaction. Work with sodium erythorbate (an isomer of sodium ascorbate, vitamin C) shows that the erythorbate accelerates the reaction. It may work by being an alternate, preferred scavenger of the oxygen, leaving the metal ions to participate in the TMAO to DMA and FA system. The data suggest the inadvisability of using ascorbate and erythorbate in fish dips for applications such as frozen gadoid fish.

The TMAOase, the enzyme that catalyses the reaction of TMAO to DMA and FA, of cod kidney lysosomes has been isolated. Antibody to this enzyme reacts with cod and red hake muscle but efforts to localize the enzyme site in the muscle have failed to date. The enzyme is also referred to as a trimethylamine N-oxide demethylase. It is TMAO-specific in certain gadoids and is inhibited by choline chloride as a competitive inhibitor. It has no methionine or tryptophan and few aromatic amino acids but does contain phospholipid which seems to be essential for activity. Its pH optimum is below 5.

The enzyme for TMA to DMA has been purified from soil bacteria. It is substrate-inhibited and can be used for a rapid assay of TMA in perchloric acid-treated fish supernatants. The perchloric acid is a protein-precipitating reagent that is added to the solution to remove proteins, which would interfere with the chemical assay.

Textural Solutions.   How can one delay, retard, or prevent the texture change? Various additives have been proposed for direct addition to fish, particularly minces, and their effectiveness is normally measured by measuring the decrease in DMA production. It was always assumed that for each mole of TMAO, one mole of DMA and one mole of FA are formed. Since the FA may react with other materials (particularly proteins) in the flesh and thus not be available for measurement, the amount of DMA produced is the appropriate measure of the effectiveness of the additive. But, again, since the FA may react after the DMA is formed, the measurement of DMA production can be misleading in evaluating a particular additive's prevention of frozen storage changes in fish.

Researchers have sought other, more direct, ways to measure the changes in fish texture to yield a better measure of the effectiveness of any treatment that delays or prevents textural change, the real problem of interest. A first need is to develop an appropriate vocabulary with which to evaluate the texture of fish minces. The words "cohesiveness" and "springiness" seem to give the best results with "raw texture panels," that is, where panelists touch the raw fish with their fingers rather than eat the product. The panelists, fish technologists familiar with the problem, reported that these terms best described the changes in fish texture that were taking place. These results were compared with those obtained from the Instron Universal Testing Machine and with those from functionality tests designed to evaluate the water-retention properties of fish muscle. A correlation was found between these simpler, quicker physical and chemical tests and the raw fish texture panels. These studies are discussed in more detail in Regenstein and Regenstein (1984) and Regenstein (1984).

Raw fish texture panels are easier to run than cooked fish texture panels, since panelists are more willing to touch and look at many more samples than they are willing to eat. It is also possible to put chemicals into the samples that people would or should never eat. Cooking itself might add changes to the samples in a nonsystematic fashion.

However, when evaluating results from raw fish samples, we must remember that the methods of freezing, storing, and thawing may have affected the texture changes, not just the fish's time in storage. Another concern is the difference between panels of trained scientists and untrained consumers. In a recent report on cooked cod that had been frozen, consumers did not perceive, or mind, the differences, but the trained panel detected increases in toughness and cold-storage flavor that were consistent with the chemical tests.

If fish minces are stored for 4 to 5 months at $-40°C$ ($-40°F$) and moved to the common United States cold storage temperatures (-20 to $-15°C$; $-4°F$ to $+5°F$), the samples no longer undergo the same negative textural changes (Samson and Regenstein 1986). This means that if the product, particularly frame minces (see Chapter 9), can be stored for some period of time at a very cold temperature, presumably in a static cold store located near the point of manufacture, then it could be transferred to the regular cold store distribution chain. We are currently confirming these results with the hope of beginning the process of determining ideal conditions for storing fish products, including fillets.

REFERENCES

Aitken, A., I. M. Mackie, J. H. Merritt, and M. L. Windsor. 1982. *Fish Handling and Processing*. Edinburgh, Scotland: Her Majesty's Stationery Office.

Hwang, K. T., and J. M. Regenstein. 1989. Protection of menhaden mince lipids from rancidity during frozen storage. *J. Food Sci.* **54**: 1120–1124.

Novikov, V. M. 1981. *Handbook of Fishery Technology*, Vol. 1. New Delhi: Amerind Publishing Co. Pvt. Ltd.

Regenstein, J. M. 1984. Protein–water interactions in muscle foods. *Proc. Rec. Meat Conf.* **37**: 44–51.

Regenstein, J. M., and C. E. Regenstein. 1984. Solubility measurements. In *Food Protein Chemistry*, J. M. Regenstein and C. E. Regenstein, eds., Appendix 27-3, pp. 325–331. New York: Academic Press.

Ryan, G. J., and J. M. Regenstein. 1985. Proper handling of frozen fish, a perishable food. *IFT Seafood Technology Group Newsletter* **4**(1): 13–14.

Samson, A., and J. M. Regenstein. 1986. Textural changes in frozen cod frame minces stored at various temperatures. *J. Food Biochem.* **10**: 259–273.

Yamaha Motor Co. Ltd. 1986. *Yamaha Fishery Journal Composite.*

8

Other Forms of Fish Preservation and Packaging

The discussions of fish preservation thus far have focused on the most current techniques which were developed to encourage the growth of fish markets in new, for example, inland, locations. Other methods of fish preservation have been around a lot longer, and in many more places around the world. Many specialty products such as cod liver and (pressed) cod roe are found far more routinely in Europe. With the exception of specific ethnic consumer demands, the United States has not been a very sophisticated marketplace for canned fish products. It is an area that should be explored more seriously.

CANNING

Canning is an excellent way to preserve and ship fish; the fish is shelf-stable and needs no refrigeration. However, at best, the quality of the final product is only as good as the quality of the fish put into the can. Thus, all of the procedures previously discussed with respect to obtaining high-quality fish must still be observed. Although a number of different species can be canned, the three most popular worldwide are salmon, tuna, and sardines in oil or water packs. Mackerel and pilchard are also popular in Europe; other experimental and unusual products are under investigation.

The Process of Canning. The heat treatment involved in canning is designed to provide commercial sterility, that is, a safe product that can be stored for a long period of time. To destroy the bacteria, sufficient heat is needed for sufficient time. Generally 120°C (250°F) is used. Achieving this temperature requires use of steam under pressure. For safety reasons, the process is carried out in a retort.

The time is determined by putting a temperature probe at the cold spot in the can along with product. The cold spot is that point in the can where heating is the slowest. For solids, where conduction heating is the main mechanism of heat transfer in the can, it is the geometric center; but for liquids, where convection heating is the main mechanism, it is approximately 1/3 of the way up from the bottom on the center axis of the can.

Based on the known rate of killing of the organism most difficult to kill, canning equations can be used to determine the heat treatment, by a time–temperature relationship, needed to leave essentially none of that organism. The higher the temperature, the shorter the time. In real products, the worst level of bacteria that might be found in foods, probably an exaggeration, is about 10 billion (1×10^{10}) bacteria per kill up to 1,000 billion (1×10^{12}) bacteria per milliliter. Once this lethality is reached, the cans or bottles are cooled.

This process is used mainly for low-acid foods, that is, products whose pH is higher than pH 4.6. For high-acid products, less heat is necessary.

The Can Itself. Traditionally cans were put together from three pieces of metal, the main body of the can and two caps. The seam of the main part of the can was soldered, often with lead. In recent years other materials are being used for solder or a two-piece can is being used. The two-piece can is made by drawing out a piece of steel so that the bottom and the sides are pushed out from a single sheet in a form-and-draw process.

The sealing of the two ends to the main body of the can involves crimping two metals together and this seal must be routinely checked to see that it has been done properly (see Figures 8.1 and 8.2). Some of the newer cans, particularly those used for sardines and other small fish, may also have aluminum tops with a pull tab and pre-scored metal to allow removal of the top without using a key.

The low profiles of many fish cans are designed to maximize heat penetration so that the fish is not overcooked. Even then, it is often possible to see some fish material burned onto the inside surface of the can, which is the hottest point in the can.

Most canned fish are species in the high-fat category and are good sources of the omega-3 fatty acids. Since the human perception of juiciness is partially a response to fat, low-fat fish are perceived as drier. They are also more sensitive to browning reactions, possibly because the browning is simply more obvious in white-fleshed fish. Taste paneling experience suggests that canned pollock and hake develop a special flavor that is quite acceptable to consumers; they were even preferred over the most expensive whitemeat tuna. Note, however, that in South Africa, preliminary trials of canning pelagic fish by adding extra fish oil led to products with unacceptable taste.

Smoking fish products prior to canning yields a very different type of product, for example, kippered herring. The color and flavor of such products usually intensify upon canning (see later).

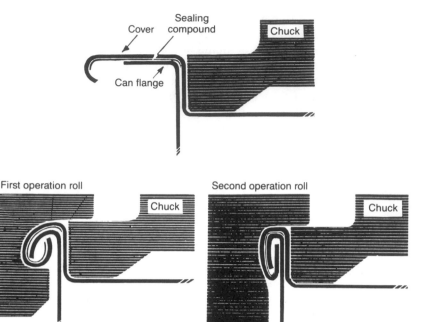

Figure 8.1   Closing a can. (*Aitken et al., 1982, p. 119; Crown copyright*)

Potential Problems

Botulism.   There have been three incidents of botulism poisoning leading to deaths in recent years because of the improper canning of commercial salmon and tuna. The most publicized incident occurred with canned salmon from Alaska. Because of Alaska's distance from the United States mainland, many of the Alaskan canneries used "folded" cans that had been shipped to them; the cans were re-formed (recircled) on site at the plants in Alaska. Apparently, the special machinery for re-forming the cans was not adjusted enough. As a result, the machinery went out of alignment and caused small pin-holes to be formed in the can. The pin-holes permitted post-canning contamination of the product by the cooling water, which contained botulism spores; this led to the botulism outbreaks. The other two incidents involved improper processing of fish, in one case from Alaska and one case from Hawaii. In at least one of these cases the problem may have been contaminated cooling water getting into the product during cooling.

Chemical Degradation.   Even properly canned and sealed fish products are at risk of experiencing specific chemical degradations. Some of these chemical

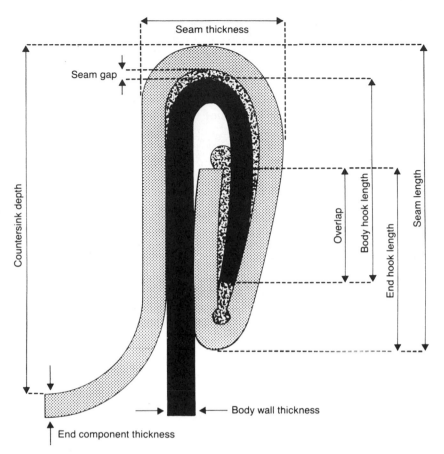

Figure 8.2 A proper can seal. (*Clucas, 1982, p. 29; Crown copyright*)

reactions may lead to smutting, that is, off-colors due to the formation of iron sulfide (FeS, black) or tin sulfide (SnS, brown). Another problem is the occasional formation of magnesium ammonium phosphate or struvite. Although the crystals are dissolvable in vinegar, struvite crystals unfortunately look and feel like glass. Their formation can be avoided or diminished by slow cooling of the canned product; adding citric acid or pyrophosphate can also help. Both smutting and struvite formation can be prevented, or at least decreased, by the use of appropriate enamel coatings on the inside of the cans.

Curd Formation.   Curd formation is another problem that sometimes occurs in canned fish. This is simply fish protein coming out of the flesh and forming a protein precipitate (gel) on the skin. Curd is particularly obvious in cooking of a fish like salmon, because the curd is white and shows up dramatically against the pink background. Curd formation can also occur with other forms of fish processing. Curd development in salmon can be decreased by a 30-second dip in a 10–15% tripolyphosphate solution. This gives a phosphate concentration of around 1%, which is twice the legal level in meats. Any ingredient added to the salmon prior to canning other than salt and/or salmon oil legally renders the product "imitation" in the United States. The international name for tripolyphosphate is sodium polyphosphate glassy. Freezing of the fish prior to canning tends to increase curd formation, as does long-term storage at low temperatures.

Stackburn.   Once again, timing is important in the preservation process. An overlong cooling period for retorted cans (possibly due to a lack of sufficient cooling water) can lead to stackburn, that is, overcooking and browning of the fish. Fish products may also require a week or two of maturing time after final processing in which the flavors in the can blend. For optimum organoleptic evaluation of the fish, the can should not be opened immediately, but rather saved and tested a week or two later. Canned fish is usually held at the plant for a short while anyway in order to identify any cans that might swell, a rare occurrence.

Canned Fish Products

Salmon.   Salmon is generally packed raw. Although it used to be broken down by hand, often by Chinese workers, a simple machine was developed many years ago to improve the process. Considered the most valuable machine in the salmon canning factory, the machine has the unfortunate and insensitive nickname of the Iron Chink. In the Alaskan canneries, the machine is usually sent back to Seattle headquarters each autumn for overhauling, checking, and safe-keeping until it is returned to Alaska in the spring.

Skinless, boneless salmon is generally manufactured in two different ways. Ralston-Purina uses a "tuna approach" (see later), doing the cleaning on the fish after cooking. Hormel uses the "salmon approach," doing the boning and skinning prior to canning of the raw fish. Canned salmon is sometimes sold as a lightly smoked product. The skinless, boneless product is not permitted by the "Standard of Identity" for salmon and is therefore being produced with a special permit from FDA.

Tuna Fish.   Tuna fish is generally cooked before canning. A very careful cooking–cooling cycle is used, but the details vary depending upon species and sizes. The cooked tuna is then separated manually into dark and light meat. The dark muscle, or "blood meat," is used as pet food because it is considered to have too strong a flavor. The lighter meat is then put into the can either manually or mechanically.

The only tuna that can be called "whitemeat tuna" in the United States is that from albacore tuna. The light tunas include a number of species, with yellowfin, skipjack, and bluefin being the most prevalent. Bonito must be sold under its own name as it is not a true tuna.

Oil, water, or a vegetable broth is added to the tuna during canning. The vegetable broth apparently aids in improving the flavor and in preventing the development of bitter notes. Some companies are now using protein hydrolysates, coming from both plant and dairy sources, as additives in canned tuna to help maintain the flavor.

The market comparisons between water-packed, with or without vegetable broth, and oil-packed tuna are interesting. Until recently, the oil-pack was generally considered superior organoleptically because it made the product richer and juicier. However, with consumers' becoming more diet conscious, there has been a shift to the lower calorie water-pack, which is also cheaper. This has opened up new opportunities for foreign packed tuna fish because in a very old treaty with Iceland, the United States agreed that the duty on oil-packed canned fish would be 35% while that on water-packed canned fish would only be 8%. The recent change in consumer preference coupled with the absence of a tariff barrier has allowed foreign-packed tuna to enter the market very successfully and has caused the demise of several American tuna canneries on the West coast that could not compete economically even with the tariff barrier.

In response to this problem, American tuna-packing of fish caught in Pacific waters now takes place in Puerto Rico and American Samoa where labor costs are lower. Yet another solution has been a "buy America" campaign by a smaller cannery that came out with a red, white, and blue label. The plan was not successful.

Sardines.   "Canned sardines" can actually be any one of many different species, the extent of the variety being controlled by individual governments. Such differences of opinion even occur within the European Economic Community (EEC), where, for example, sprat can be packed as sardines in all the member countries but the United Kingdom. All United States sardines are packed in Maine. The product is the tail portion of herring cut behind the anus. The front portion of herring is generally used for lobster bait. Packing sardines is a manual-labor intensive operation.

American and Canadian sardines are canned in water, oil, or sauce. Recent efforts to offer a wider and more interesting variety of these sauces has helped create some new market interest in this product. Consumers can now choose from among traditional mustard or tomato sauces and, for example, newer hot sauce, barbecue sauce, or lemon sauce. Large herring are now being cut like "steaks," creating yet another variety of canned sardine product. All of these products are generally steamed while in the can; if necessary, the cans may be turned over to allow for the removal of any liquid generated at this time. Then the packing liquid is added and the cans are closed. American sardines are not peeled, although this is a common practice in certain countries such as Portugal.

Pilchard.   Pilchard, a closely related species to sardines and tuna, used to be canned mainly in a tomato sauce in most countries. More recently, some has been packed in a form similar to tuna, especially by canneries in Peru and Chile, and has been sold at a very low price. However, with a decrease in stock availability because of El Niño (the warming current off of South America) and a price decrease in tuna, pilchard has all but disappeared, at least temporarily, from the United States market.

Mackerel.   Canned jack mackerel, packed either in California or overseas (e.g., Chile, Thailand, Japan) is a wonderful secret in the American marketplace. One-third the price of sardines ($0.89 per one-pound can in 1989), it is a healthful product that is utterly misunderstood by the average consumer. Unfortunately, the picture of a whole fish on the package does not help. Are the companies suggesting that a whole fish is actually inside the can? Or that this product should be prepared like a fresh fillet? The consumer is likely to be confused. In fact, the can holds a headed and gutted fish which can be used very effectively in many tuna and salmon recipes. Its distinctive fishy taste also makes it a good substitute for herring. Consumer education on this product could have great benefits both to consumers and to the fish industry. Unfortunately, the East Coast counterpart of this fish is currently assigned to a directed foreign fishery because Americans think mackerel is only used for bait.

At recent fish training workshops for New York State Extension Agents and others dealing with the public, we have prepared a "Poorman's Herring Salad," a cold salad using a recipe designed for sardines. We did a preference comparison and found that the product made from mackerel was preferred two to one over that made with sardines.

Canning in Jars

Some products are canned in jars. Many of the pickled products use high acidity as a preservative (see later) and are not heated. However, some products are heat-cooked to give commercial sterility. It is important that products in a jar, which are usually processed in a water-retort system, be warmer than the water temperature until after the retort is pressurized, to prevent contaminated water from entering into the product.

One of the canned products that is usually glass-packed is gefilte fish, a Jewish delicacy that used to be prepared by homemakers with pike, carp, and whitefish, numerous hours, and much effort (see Chapter 13). In the United States various freshwater white-fleshed fish are ground together and formed into an elongated fish ball. This is usually extruded into a 82°C (180°F) water bath to set the product and then it is carefully (manually) packed in jars and retorted. This product is generally presented in a clear or gel broth and requires very careful retort control to ensure sufficient heat for safety with regard to botulism. Overheating at processing conditions slightly beyond the FDA-approved safe level causes browning. Large cans and bottles yield a darker product for the same heat treatment.

A final word concerning canned fish. Many of these products are packed complete with the fish bones. The high temperatures and/or pressures reached during canning soften these bones until they are quite soft and easy to eat. Bones are an excellent source of calcium, one of the nutrients that is often deficient in the American diet. Recent research suggests that it is not only important for bone development, but that it may also help reduce heart disease and hypertension. Post-menopausal woman need extra calcium to lessen the brittleness of their bones. Nevertheless, it is very difficult to convince people to eat these bones. We recently had the opportunity to serve two versions of "Poorman's Herring Salad" at a reception where we were able to test whether a bone-in or bone-out product was preferred. The product with the bone-in, the first author's preference, was preferred by 34 of the people sampling the product; the product in which the bones and fin material of the canned mackerel were removed, the second author's preference, was preferred by 31 people. The reason given by many of the people for preferring the bone-in product was a stronger flavor, probably attributable to the bones and the marrow. Only one person commented that she had noticed the bone in one sample and therefore preferred the other. We recommend carefully mashing the softened bones between the fingers during preparation.

SMOKING

Ancient methods of smoking fish were used to preserve the food product. Modern smoking is more generally done to meet consumers' organoleptic requirements. The process takes anywhere from a few hours to a few days. In most cases, the product must still be refrigerated to survive on the shelf; and even then, the smoked fish may not last much longer than does fresh fish. Experimental samples with a mixture of citrate, salt, and sorbate protected the finished smoked fish for up to 4 weeks at room temperature; the control samples lasted less than 1 week. The addition of BHT, a synthetic antioxidant, to the brine extended the ultimate shelf-life for another 4 weeks. The spoilage of smoked fish is often due to molds. In general, smoked fish products freeze very well.

Fatty fish lend themselves to smoking more readily than low-fat fish. Some popular products include various types of smoked herrings, bloaters (made from ungutted fish!) and kippers, a herring smoked with the brown dye "FK," which is illegal in the United States. In the spring, there is also a "spring oily herring" which can be smoked. Because these fish have usually been feeding so voraciously, the oil from their own food is not yet absorbed into the fish's flesh.

Brining.   The first step in smoking fish (whether gutted, headed and gutted, or skin-on fillets) is usually salting (brining) the fish with a 70–80% brine. A 100% salt brine sometimes causes the salt to crystallize; a 90% brine causes salt-burn. A lower percentage brine, 50% or so, causes swelling because the fish absorbs water from the salt solution; excess moisture will eventually have to be removed during the drying stage of smoking. Some smokers will use a dry salting, that is, direct addition of salt. The drying of fish before hot-smoking helps to prevent the fish's breaking up into pieces during "cooking."

The fish's fat and skin brine more slowly than the flesh itself. Contrary to popular belief, the fish must be of high quality initially if it is to be of top quality when smoked; smoking is definitely not a way to save "old" fish. The gloss associated with smoked fish products is due to the extraction and setting of protein at the surface. Older, that is, longer postmortem, fish do not yield this protein extraction and there is no gloss.

Brining before smoking serves a number of purposes: (1) it firms the product; (2) it stops the oil from oozing out; (3) it imparts flavor; and (4) it prepares the flesh to accept smoke. The increased salt level inhibits some of the food-poisoning microbes. In some countries, color is added to the brine. For example, in the United Kingdom cod is often colored yellow while kippered herring is always brown.

The product is then hung for smoking on tethers (hooks, or—more often—rusty nails!) or on smoking racks. The latter generally cause markings on the fish skin.

The Smoke Itself.   Smoke may be generated by little fires in the room (smoke house) (Figure 8.3) or, in the case of a modern kiln, by smoke from a nearby smoke generator. The latter devices involve some type of heating element to generate smoke at temperatures in the 250 to 350°C (480–660°F)

Figure 8.3 A traditional kiln. (*Aitken et al., 1982, p. 101; Crown copyright*)

range. Temperatures above 400°C (750°F) cause the formation of undesirable compounds suspected of being carcinogenic.

Different smoking woods impart different characteristic colors, flavors, and aromas to the final product. Softer woods color and flavor the product faster.

Some of the cures (brines), used with fish include nitrite, which gives some protection against botulism. However, there is concern that the addition of nitrite to fish, which has high levels of small nitrogenous compounds such as free amino acids, TMAO, and TMA, might lead to the formation of carcinogenic nitrosoamines.

The smoke itself is made up of two fractions: the tarry droplet fraction contains polycyclic aromatic hydrocarbons which have also been identified as potentially carcinogenic; and the volatile fraction contains the flavor compounds, phenols, and carbonyls. The latter can be more easily duplicated by liquid smoke, which adds the advantage of eliminating the questionable tarry fraction. Liquid smoke is beginning to see an increase in use in the United States, especially for salmon. Recent research by M. Eklund and his colleagues at the NMFS Seattle laboratory has shown liquid smoke to possess some antibotulinum activity.

Many liquid smokes used in the United States are prepared by a controlled combustion followed by a water extraction of the smoke vapors. According to the NMFS in Seattle, these products can inhibit botulism in a product with a 1.5% water phase salt (see next paragraph); more than 3.7% salt was needed in the traditionally smoked controls. Some recent products have used oil extractions of the aqueous phase; the inhibitory salt level varies depending on the particular material.

Fat and proteins do not dissolve salt. Therefore, the actual level of salt in a product with respect to its ability to prevent bacterial growth depends on the concentration in the aqueous phase. This is measured and reported as the water phase salt. It will always be higher than the salt concentration measured on the product. This should be measured in the center of the thickest part of the product, which is where diffusion will be slowest.

Regardless of the type of smoke used, it is the smoke that is moved past the stationary fish. The air flow must be controlled carefully. Even in modern smoke houses, it is often necessary to take out all the fish half-way through the smoking period and rotate the carts or hanging rods so that all of the product is smoked evenly.

Drying.   Effective smoking must be accompanied by sufficient drying of the fish. As always the rate of drying, or lowering of the A_w (water activity), depends on the humidity and the air movement. The water activity is the way technologists measure the amount of water that is "available" in a product. It is sometimes called "free" water and sometimes referred to as the water that is available for bacteria and other microorganisms to grow on. Technically it is the same as the relative humidity of the air surrounding the product under carefully defined equilibrium conditions in the laboratory. This is not the humidity in a real room.

Generally a water activity of 0.85 (85% relative humidity) or higher will permit many different bacteria to grow. However, the histamine-forming bacteria, for example, do not generally grow below an A_w of 0.90. A much lower water activity, probably about 0.65, is needed to prevent the growth of yeasts and molds. Visible mold growth occurs at $A_w = 0.85$ in 4 days, at $A_w = 0.80$ in 10 days, at $A_w = 0.75$ in 40 days, and at $A_w = 0.75$ in 100 days. Aflatoxins, products of molds that are considered quite toxic, are not produced by molds if the A_w is below 0.85.

Some products depend on their low water activity to obtain their preservative effect. A few examples are jams, honey, nuts, and pasta.

Humidity rises with time in a closed system. On warmer days, the increase in water-carrying capacity of the air decreases between the amount of heat applied (temperature increase) to the incoming make-up (new) air is less. The smoke house operator must therefore regulate the amount of air released from the smoke house carefully: too much air released, and heat and smoke are wasted; too little air released, and drying literally takes forever. According to a recent report from the University of Washington, a microwave oven can be used in conjunction with smoking to speed up the drying process.

The drying process strengthens the fish and helps it to survive the cooking process without falling apart, particularly if the fish have been hung. Drying aids in the formation of the glossy surface pellicle by sealing the surface. It also helps avoid coagulation of the white curd, protein, that sometimes oozes out of the fish, and case hardening, that is, an excessive drying of the surface which prevents the moisture of the product internally from escaping or the smoke from getting in. The curd is an excellent medium for bacterial spoilage.

Smoking can be done in either of two distinctive temperature ranges. Having received the previously described heat treatments, hot-smoked fish is a fully cooked product. Cold smoking, however, involves keeping the smoking temperature below 30°C (86°F), or 40°C (104°F) for tropical water fish, throughout the smoking process. This process is used for most of the smoked salmon made in the United States, including lox. However, the final product does not have an extended shelf-life since to provide that would require an 8% water phase salt level—clearly stronger than most of us would care to eat. During smoking, an initial humidity as high as 90% maximizes smoke adsorption; an eventual humidity of 70% seems to be a good compromise between maximum smoke adsorption and minimum case hardening. In the United States, cold smoked salmon is usually made from frozen salmon, thereby eliminating any potential health problems from parasites.

For any smoked fish product, the following temperature/water phase salt levels are recommended by the FDA to prevent botulism:

5% NaCl, 65°C (149°F)
3.5–5% NaCl, 82°C (180°F)

Unfortunately, these recommendations have not been widely accepted by the United States smoking industry because of perceived organoleptic deficiencies. On the other hand, the United States government has discouraged the vacuum packaging of fresh, but not frozen, smoked fish until the above guidelines have been met. The situation is currently unresolved.

STOCKFISH (DRIED FISH)

When fish is simply dried, the resulting product is often referred to as stockfish. In some areas of the world, particularly Scandinavia, it is sometimes possible to dry the fish naturally in a relatively cold environment so that a very fine product is obtained; the product can be eaten raw. Natural winds provide the required high air flow. However, when other fish are sun-dried in more tropical conditions, the quality is more difficult to control—especially if the product is dried on the ground (often on sand). Blowflies can invade the product so easily that in some countries it is estimated that as much as 20% of the protein content of the "dried fish" is actually derived from insects. Dryers are being designed and built to improve dried fish quality around the world, especially in many Third World, tropical countries. These dryers may incorporate the addition of heat (by fire or the better use of solar heating). The importance of a proper air flow pattern in the dryer cannot be overemphasized.

Solar heaters or dryers are often distinguished by whether they use natural convection or forced convection to obtain the necessary air flow. In the former case, the design of the dryer is such that air flows are provided naturally, while in the latter case there is a provision for fans or other means of mechanically moving the air. Although they are more expensive, forced air flow dryers usually have a higher capacity per unit volume.

Drying of any product usually occurs in two steps. During the first step, the constant rate drying period, the air flow is critical. There are always two layers of air between the fish and the moving air. Closest to the fish is a thin stationary air layer; the second is a slow-moving air layer. The amount of moisture transfer from the fish to the air also depends on the relative humidity: the lower the humidity, the better the moisture transfer. However, if the rate of water removal is too high in comparison to the diffusion rate (see later), then case hardening occurs.

The second step of the drying process is the falling rate period. In this step, increased temperature has a greater effect on the rate of drying than does the air flow. The rate-limiting process in this step is the rate of diffusion of the water to the surface of the fish. Fat interferes with the diffusion of water to the surface. It is a challenge to balance considerations of drying rates and quality/safety issues.

Other commercial dried products use only a part of the fish, for example, fish heads. Popular in Nigeria, their importation was dependent on the country's short-lived economic success in exporting oil. However, many other fish products are dried in Africa, both on the ground and hanging. These products are often grilled at the same time that they are dried. Moisture content may vary from a safe 10% to a more perishable 60%.

Lovers of Oriental food are probably familiar with shark fins. This specialty dried product is generally air-dried at sea. The shark fin is high in collagen (gelatin) and develops a fishy taste during drying. The material is sometimes broken down into fibers and sold "shredded." Shark fins are a highly prized item in the Orient and enjoy a thriving market. Typically quality and grade standards are quite complex in the Oriental fish market to satisfy a wide range of consumer needs and attitudes.

FREEZE-DRYING

Most commonly used for vegetables and coffee in the United States, the technology of freeze-drying has found limited application with fish. The major use of the process is for products that must be light in weight and easily rehydrated such as "instant" soup products. However, freeze-drying has seen only limited use for such instant products because it essentially creates an airy sponge that is easily oxidized. Thus, rancidity can be a problem unless the material is vacuum- or nitrogen-packed. When fish is properly freeze-dried, the resulting product shows no shrinkage and no case hardening because the proteins are not subjected to any thermal damage. The freeze-dried product rehydrates fairly quickly but does not retain the reabsorbed moisture on cooking.

SALTING

The Salting Process. The process of salting or "brining" fish for preservation purposes is an ancient one. It is generally followed by drying and/or smoking. However, its purpose is not just to preserve the fish; it is to limit any possible spoilage in the entire process by reducing the water activity.

Salting can be done in a number of different ways. The first is called dry or "kench" salting. For a light salting, the fish is mixed with one part of salt for each eight parts of fish; for heavy salting the product is mixed with one part salt for each three parts of fish. The fish is then piled up on a clean surface in piles of two meters each. The head end of the fish is higher than the edges. Fish are thicker at the nape than the tail, even as fillets. The liquid that is pressed out of the fish is allowed to run off, that is, it follows the natural slope of the fish. The product is then restacked every few days until the proper salt penetration and moisture loss is obtained.

A second way to salt fish is to brine them. A saturated brine, rather than dry salt, is generally used for this purpose. The purpose of the closed brining method of salting fish is that it keeps air out, thus limiting the amount of rancidity that can develop. It is the preferred method for salting fatty fish. In any case, if the brine is less than 12% saturated, the fish absorbs the brine; if the brine is above 12%, the fish loses water and solubles.

A third method is to combine the salting and brining procedures described above. The fish is salted dry but then placed in a barrel so that the liquid that comes out of the fish can be collected, thereby forming a brine. In this system, saturated brine usually has to be added to "top off" the barrel after a few days. (360 g of NaCl per liter of water gives a 100% saturated salt solution.)

Once a fish has been salted completely by any of these methods, the fish can be dried. Sun drying is often used in cooler climates; in warmer places, solar or heat dryers are more common. The final product must have a water activity (humidity) below 0.75 (75%), that is, the product must have a composition of about 35% water and 28% salt.

Salted Fish Products.   A few special salt products deserve particular mention. Saltfish is made by leaving fish in salt for 3 weeks. It is equivalent to 1.7 times its weight of fresh fish. It keeps best at 2–4°C (35 to 40°F). Dried salted codfish (klipfish) is salt fish that has been washed, recleaned, salted and dried, often in a hot-air drying plant. It is equivalent to 3.2 times its weight of fresh fish and keeps well at around 5°C (41°F). Recent experiments in Iceland have focused on a new process in which cod is salted only once instead of the traditional twice, supposedly yielding a lighter, fresher fish taste.

Headed and gutted, washed, fresh sardines are cured in a saturated brine solution for 6 days, then drained and pressed at 465 kg/m^2 for 15 hours. They are stored at 28°C (82°F) and have a shelf-life of about 4 weeks, longer with lower fat sardines. The final water activity is about 0.78.

Challenges.   There are problems to be solved when salting fish: first, of course, is the quality of the salt itself. Sea salt is often quite dirty—both microbiologically and in terms of its trace mineral content—and cannot be used without special recrystallization procedures.

Another problem with salted fish is mold and bacterial spoilage: either the pink appearance due to halophilic bacteria or the Dun which is caused by the brown mold *Sporendonema epizoum*. The latter seems to be more prevalent in rock salt. Sorbic acid inhibits the growth of halophilic organisms (pink) in salted fish; however, its use is disallowed in countries under French law. In these cases, sulfur dioxide is permitted, although it is not as effective in inhibiting the growth of the halophiles.

The halophiles in salt are quite hardy. Temperatures as high as 160–180°C

(320–356°F) may be needed to kill these organisms. Irradiation is ineffective because it takes so long that the salt becomes discolored. Cooking the salt in an oil bath is effective, as is using a microwave oven for 10–15 minutes.

On the other hand, low temperatures, below 10°C (50°F), help preserve salted fish. Sometimes a preservative such as potassium sorbate or sodium propionate at a 3% level can be added to inhibit mold growth. Metabisulfide can also be used for this purpose at times.

If fish weigh more than 35 grams, one cannot use salt as a preservative at room temperature. For example, in the tropics, the fish would spoil before the salt was able to penetrate.

The sodium chloride itself leads to a slight softening and yellowing of the fish. A small amount of calcium and/or magnesium salt can help make the sample whiter. Levels of 0.15–0.35% calcium are recommended. Higher levels seem to slow the penetration of sodium chloride. Some researchers are trying to take advantage of these interactions between salts to control the amount of salt taken up by fishery products, especially with smoked or brined products. The magnesium level must be kept below 0.15% or a bitter flavor results.

Trace minerals such as copper must be kept below 0.1 ppm (parts per million) and iron below 10 ppm if a brown discoloration is to be avoided. This may be difficult with sea salt.

If the surface of the fish dries out too quickly, salt penetration is slowed. This phenomenon, similar to case-hardening, is called "salt-burn" and is caused by using too small a salt-grain size. In this case, the salt dissolves too quickly and the fish's insides eventually spoil, yielding a "putty" fish.

Before salt fish is consumed, the general practice is to soak the fish in several changes of water and to discard the water. The fish is often cooked once or twice in water and the water is discarded again. With proper washing, the final salt level is generally no higher than that found in many other processed food products.

Salted fish has a distinctive and characteristic flavor that is actually due to a mild rancidity. Remarkably and unfortunately, this rancidity becomes a flavor expectation for fish in some countries.

PICKLING (MARINATING)

The pickling process resembles the brine salting process except that vinegar is generally used as the preservative instead of salt. The resulting product's flavor is determined by the actual composition (sugar, salt, spices, etc.) of the pickling medium. The product is usually consumed raw, with the pickling softening any bones present so that they can be eaten.

Pickling is popular in Scandinavia, especially for herring, which is sold in the United States in bottles produced either there or abroad. The first soak in 4–7% vinegar can last as long as 3 weeks and is referred to as the "finishing" bath. This process softens the bones. Following this step, the herring is generally repacked in 1–2% acetic acid and 2–4% salt and sold in this solution. Other acids might also be used. The blend of spices often becomes an important company secret. One variation uses sour cream instead of the second pickling brine. The sour cream must be carefully made so that it remains stable and does not curdle in the high-acidity environment. These products must all be stored in a refrigerator, even before the bottle has been opened. In recent years, imported pickled herring from Scandinavia and other countries has become more readily available in United States markets.

Another pickling process for herring uses salt for the first cure and yields a much firmer fish. This product is more popular in the Jewish trade. It is sold in the northeast of the United States as compared to the Scandinavian style which is popular in the American midwest.

Another type of pickled herring, popular in Europe, is referred to as "Matjes" herring. This is usually made from small fatty herring which are lightly salted and matured with special seasonings for about 8 weeks.

FISH PASTES AND OTHER FERMENTED PRODUCTS

Fish is fermented into pastes in many of the Oriental countries. This process generally involves a complete or almost complete hydrolysis of the fish using their own digestive (proteolytic) enzymes. The process preserves fish and provides a very distinctive flavoring material. The products are usually quite pungent and salty since added salt is often used to preserve the product while postmortem digestion takes place. Loc Moc from Vietnam is a typical example.

A product that does not quite fit into any of the above categories is Pindang from Indonesia. Made by boiling fish in a saturated salt solution, it is then sold as is and lasts for 3 to 6 days without refrigeration.

A process for producing various fermented products in Africa has been described. Senegalese guedj is made from split, washed fish. These are lightly salted and fermented for a few days in wooden tubs. The fish are then dried on racks. The bacteria (mostly coryneform) for the process are assumed to come from the wooden tubs.

Tambadian is prepared from whole fish that are buried under hot sand for a day or two during which time autolysis takes place. Those with bloated bellies are gutted, split, salted, and then sun-dried for a few days. In Ghana a similar product is produced from already-spoiled fish by putting salt in the belly cavity, fermenting the fish in a basin for a few days, and then lightly sun-drying them.

IRRADIATION

Irradiation using either cobalt-60 or gamma rays can be used to preserve various foods, including fish. The food is not rendered radioactive. It is not clear whether the United States FDA will define this technique to be a form of processing or whether it will be considered the equivalent of a food additive. In the latter case, irradiation would have to be declared specifically on each product label.

Appropriate doses of irradiation make possible a room-temperature, shelf-stable product. However, even under these circumstances, the process does not stop the natural fish biochemistry. Radiation apparently accelerates rancidity development in fatty fish. It is also not clear what effect a high-dosage sterilization treatment would have on fish quality; the safety of these levels has not been fully established to the satisfaction of the FDA. Consumers would probably be unhappy to learn that some imported shrimps that have been rejected for salmonella in the United States have been known to be shipped back to Holland, irradiated, and returned to the U.S. marketplace.

Radiation-sterilized cod fillets will turn brown over a short period of time. Breaded fish products that have been prepared as a shelf-stable irradiated product will have a texture typical of that associated with the long-term storage of gadoids. Two-month old red hake stored at 0°C (32°F) also showed signs of gadoid texture change. These early experiences suggest that shelf-stable irradiated product would undergo undesirable texture changes because the enzymes (like those that take TMAO to DMA and FA) would have enough time and an appropriate temperature to cause damage even if the reaction rate remained low.

Because fish are currently caught and handled in widely dispersed areas, and because the industry already has problems maintaining high quality of fish, it is not clear whether irradiation will be of much benefit with respect to quality even though it can prevent some amount of spoilage.

RETORT POUCHES

"Commercially sterilized" retort pouches have the advantages of packages that can conform almost perfectly to the shape of the product. The pouch concept is meant to decrease the profile of the product being heated so that the distance from the edge to the center is minimized. This permits the same commercial sterilization effect on the product as with regular cans, but in a noticeably shorter period of time for the heat treatment. In addition to potentially yielding higher-quality canned fish, this type of packaging could also enhance the packing of whole fish, whole steaks, and/or whole fillets.

Shelf-stable products are generally reheated the same way as boil-in-bag products (which require refrigeration). Another method is to remove the product from the pouch and cook it in a pot like a canned product. Neither method has satisfied American consumers. Treating the retort like a canned product leads to loss of integrity of the large fish pieces at a higher price. Leaving the product in the pouch during cooking has met with the same rejection as boil-in-bag products. A relatively acceptable product can be made by retorting whole pieces of salmon. It is not as "cooked" as canned salmon and the product will retain its shape.

ACIDIFICATION

The acidification and salting of fish flesh leads to various changes in the material. The lowering of the pH generally leads to a loss of the fish system's water-retention properties and consequently to a toughening of the flesh. However, acid may aid in the removal of fat, and may also lead to an eventual softening of the fish by its effect on the solid components of the flesh.

REFERENCES

Aitken, A., I. M. Mackie, J. H. Merritt, and M. L. Windsor. 1982. *Fish Handling and Processing.* Edinburgh, Scotland: Her Majesty's Stationery Office.

Regenstein, Joe M. and Carrie E. Regenstein. 1982. *Choose Your Title: Kosher Minced Fish Cooking; Fish Cooking with a Food Processor; International Fish Recipes.* Albany, NY: New York Sea Grant Institute.

Clucas, I. J. 1982. *Fish Handling, Preservation and Processing in the Tropics,* Part 2. London: Tropical Products Institute.

9

Special Processing Procedures

A number of special processing procedures have been sought and developed to increase the usability of fish as a human food. Some represent ethnic taste preferences; others are a deliberate attempt to utilize wasted resources to feed a hungry world. Pig farmers reputedly encourage market use of "everything but the oink"; fish are silent, but the methods described in this chapter should help encourage use of absolutely every part of a given fish and help plug a leaky fish pipeline (Figure 9.1).

MINCED FISH

Minced fish is a hamburger-like product that is usually created by mechanically grinding up and then removing the edible flesh from parts of fish that are otherwise thrown away. The idea of mincing fish for human food came about as a result of several problems: (1) filleting fish leaves a lot of edible fish on the skeleton; (2) too many species of fish are not accepted in the marketplace, particularly if they are boney, and are therefore underutilized; and (3) too many small fish are thrown back into the water, often dead. Processing methods previously described in this book have not been sufficient to solve these problems.

Understanding the anatomical makeup of a given fish clarifies the desirability of using the mincing process as a solution. For example, these estimates might represent cod: the guts (entrails) generally constitute about 16% of the fish's total weight. For many of the major commercial species, these are removed at sea. Of the landed weight of the fish, the fillets as initially cut are about 47% of the weight. The skin removed from these fillets accounts for about 3% of the landed weight. Creating a boneless fillet by removing the pin bones (with a V or J cut) generally accounts for another 5–7% loss of the original weight. The material removed as the V cut is a very high-quality fillet material. Yield can be significantly increased if the loss for V cuts is reduced. Water knives or laser cutting have been suggested

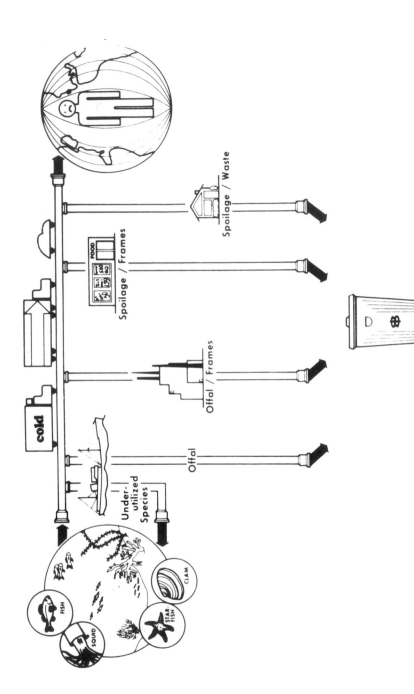

Figure 9.1 The food from the sea pipeline

to make the bone removal cut more exact, that is, closer to the line of bones. These two types of cutting equipment are supposed to be "clean" pieces of equipment. In fact, a water-jet operated by a computer for portioning has recently been announced by a company in the state of Washington.

The remaining rack, skeleton, or frame with the head still attached constitutes about 40% of the landed weight: about 25% is the head, and 6% is the backbone. A new machine under development in Iceland (by Kvikk) is supposed to remove about 7–15% of the landed weight as head meat mince. Their new cod head splitter enables processing of the edible parts of a cod head, giving a claimed 5–13% yield improvement. The fish must be 55 cm to 110 cm (21 in to 443 in); the rate is 30 to 35 heads per minute. A Canadian machine by Abco removes the "cod tongue" from fish heads. These pieces can be battered and fried. Such products contain a lot of connective tissue (collagen), giving them a unique texture. The product is considered a delicacy in Newfoundland, but Americans may find it an acquired taste.

To use the underutilized part of a fish as a human food, the edible, nonfillet parts of the fish are run through a "deboning" machine. For example, the rack of the cod described earlier could be put into the "deboner." If the backbone has been removed, this would be starting with about 22% of the total landed weight of the fish. The result would be about 15% frame mince and 7% waste of the original landed weight of fish.

Deboning Machines.   There are a few different types of deboning machines. The first is the belt-and-drum deboner, first developed in Japan, in which the belt and the drum usually run at the same speed (Figure 9.2). The flesh, blood, and fat are much softer than the bones, scales, and skin. Therefore, when the fish is pressed between the drum and the belt, the former go through the holes of the drum and the latter do not; the materials are thereby separated into two fractions. The softer material that has been pressed through to the inside of the drum is called mince, and might be thought of as a fish "hamburger" meat or ground fish. Most belt-and-drum deboners have holes of 3 or 5 mm ($\frac{1}{8}$ in or $\frac{1}{5}$ in) for use with fish: 3 mm is probably the most common size because it yields a more textured final product rather than a smaller hole, which gives a "pasty" product. A larger hole may allow an unacceptable level of bone through the holes. Pastiness is the goal when shellfish carcasses, particularly crustaceans such as crabs and lobsters, are run through the machine to create flavoring agents for other products such as surimi-based crabs; a hole size of about 1 mm works well in these cases. The machine itself can accommodate a hole size as large as 10 mm (0.4 in).

The second type of deboner is the auger-screen machine, in which an auger rubs along the inside of a drum with much smaller holes (approximately 0.5 to 1.5 mm) to move and crush the material (Figure 9.3). Newer equipment runs with higher auger speeds and a small gap between the auger and the

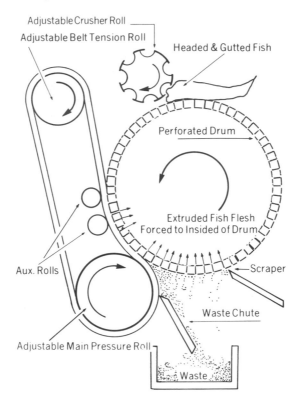

Adjustable Crusher Roll
Adjustable Belt Tension Roll
Headed & Gutted Fish
Perforated Drum
Extruded Fish Flesh
Forced to Inside of Drum
Aux. Rolls
Scraper
Waste Chute
Adjustable Main Pressure Roll
Waste

Figure 9.2 A belt-and-drum deboner. (*Lanier and Thomas 1978*)

drum. The smaller hole size is necessary to exclude the bone since the pressure generated in these machines is greater than that found in belt-and-drum deboners. Yield is higher from auger-screen deboners, but the final product is always quite pasty. To maximize yield, it might be desirable in some circumstances to use a belt-and-drum deboner at low pressure to yield the highest quality material and then to take its waste stream through an auger-screen deboner to capture the lower-quality material. If the waste stream from mincing is to be used in fish meal or other by-products (Chapter 10), then the belt-and-drum deboner might be run at an even lower pressure setting so that some protein is deliberately left with the waste stream.

A third type of deboner uses a hydraulic ram to squeeze the flesh past a screen in an automatic batch-type system. The advantage of this machine is a decrease in the heat build-up in the fish flesh, which can be a problem with the other two types of deboners. With enough back pressure, it is possible to cook the meat coming out of the auger-screen deboners.

Companies making the belt-and-drum equipment include Baader, Bibun,

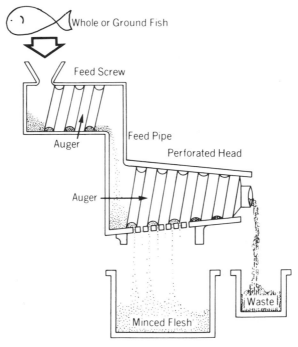

Figure 9.3   An auger-screen deboner. (*Lanier and Thomas 1978*)

and Yanighiya; Beehive and Yieldmaster make the auger-screen machines; and the Protecon is an example of the hydraulic-ram type of equipment. The latter two types of deboners are stronger, that is, they have no "rubber" belt to wear out; thus they are the machines of choice for poultry and red meats, which have much harder bones. The belt-and-drum deboners are most popular for use with fish and shellfish.

Processing Minced Fish.   A useful scale for mince texture measurements in fish was developed for hake by the government fish laboratory in South Africa:

Texture score sheet for hake*

1. Extremely soft, mushy, pasty, totally lacking minced fish texture
2. Very soft, slightly mushy, lacking minced fish texture
3. Soft
4. Firm, neither soft nor tough, typical fresh minced hake texture

---

* Courtesy of the South African Fish Industry Laboratory.

5. Slightly tough
6. Tough, fibrous
7. Very tough, rubbery, crumbly, stringy

The same experiments used to create this score sheet also showed the significant improvement in hake mince texture and general quality at $-25°C$ ($-13°F$) versus $-18°C$ ($-0.4°F$) frozen storage. The effect of various additives was much less distinct.

Note that mincing of hake causes more rapid denaturation, for example, the loss of solubility, than with other species. Although the free fatty acids increase during storage, mincing is not necessary to accelerate the (phospho-)lipases. Apparently, both protein denaturation and free fatty acids are involved in the observed changes during frozen storage.

Minced fish, particularly frame mince ("otoshimi" in Japanese) has handling and storage problems. This is especially true if the mince is derived from gadoid fish. The mixing and grinding inherent to the mincing process cause the material to undergo the standard gadoid texture changes found during frozen storage much more rapidly than do regular fillets. Thus, frame mince held at -14 to $-20°C$ (7 to $-4°F$) for a month or so almost certainly picks up an undesirable spongy, cottony texture. The incorporation of air that always occurs during mincing inhibits the rate of this reaction, but apparently not enough to offset the other effects. Preliminary research suggests that using a cold enough temperature, that is, around $-40°C$ ($-40°F$), allows the mince to be stored successfully for a few months. Apparently, stabilizing changes occur at that temperature that prevents subsequent texture deterioration at higher cold-storage temperatures. In such a case, the mince could be released into the more traditional North American cold-store chain with minimal further damage. Thus, if appropriate extra-cold stores were built for storing the mince until it was conditioned (about 4–6 months), then it would not be as necessary to change the cold-store distribution chain temperature of -5 to $+10°C$ (23 to $59°F$) in the United States. Remember that freezing facilities require much more energy input than cold stores. The two types of facilities and their purposes should always be carefully distinguished.

Liquid Carbonics has recently announced a stored energy carbon dioxide system. By using the triple point, the unique point where the gas, liquid, and solid can exist together at the same time, of carbon dioxide (approximately $-56.7°C$, $-70°F$; 60 psig) in a closed refrigeration system, cold as solid dry ice can be generated and stored at cheaper evening rates and then used during the day. This type of system saves on straight electrical costs and permits a decrease in electrical "demand" charges. When purchasing commercial or business electricity, one not only pays for the amount actually used, but also for the maximum amount of electricity available during prime hours. The latter is the "demand" charge. It usually does not apply to residential electrical rates. This system should allow for significant cold-storage savings for the extra-cold temperatures required by fish. As the frozen food industry has been developing more up-market fish products, particularly entrees, it becomes increasingly important to insure that the products are of excellent quality. The time may be appropriate for this type of extra-cold storage to be used with fish products.

Another way to stabilize fish minces is to add appropriate cryoprotectant agents such as starch, which may be an ingredient of the final formulation anyway. Sugar and sugarlike substances such as sorbitol might also be used; they have been successfully used with surimi. Other compounds have been proposed, for example, polydextrose (which does not contribute to the product's sweetness), egg white, and dried potato. Alaskan pollock mince products made with 5% egg white or 3% dry potato evaluated in the form of a battered and breaded, deep-fried product were preferred by a taste panel over minced products made with sodium tripolyphosphate.

Minced Fish Products.   What can be done with minced products? Mechanically deboned poultry is the major ingredient in chicken hot dogs and many other further processed chicken products. It is usually obtained from the backs and necks of poultry that has been cut into parts and from poultry carcass waste, for example, breast bones after meat has been cut away (manually) for further processing, that is, as boneless breasts.

The red meat equivalent is required to be called "mechanically processed species (e.g., beef, pork) product" (MPSP). It has been tied up in regulatory knots in the United States Department of Agriculture for years. The hand-trimmed bones which would be the raw material for MPSP are currently being used for lower-value processes, such as rendering. This represents a yield loss of about 30 lb of edible flesh per steer.

Some of the minced fish is used for traditional products such as sticks or fingers, portions, and cakes. In certain other countries, a high-quality white mince, for example, "V-cut," representing 10–20% of the total weight, may be combined with fillets in the manufacture of fish blocks. This works very well because the mince fills in many of the voids or air pockets in the blocks, and also helps to bind the block together. Currently pending is a United States standard for just such a product. The standard also includes a proposed methodology for measuring the amount of mince that has been mixed in with the fillets. The National Marine Fisheries Service is currently testing the method with a collaborative study.

An earlier report from the NMFS Gloucester lab suggested that the addition of mince to fillet blocks, up to 30% by weight, did cause a change in some of the chemical parameters; however, the sensory results did not change.

Mince can also be added to red meats. The NMFS petition dates back to 1980, and is based on work at a private laboratory (ABC Laboratories) in Gainesville that showed that the fish product with more protein and less fat was acceptable to consumers. This fact could lower the cost of the meat-block ingredients, and would offer consumers another choice. However, the United States Department of Agriculture's Food Safety and Inspection Service (FSIS) required more information on the species of fish involved; the chemical composition, microbiology and toxicology of each species involved; information on fish inspection; grading and standards for minced fish; information on the processing of minced fish; sanitation; bone fragments in the final product; nitrite studies involving fish; nutritional data; and consumer surveys dealing with the consumers' perceptions with respect to labeling. It is no wonder that the regulatory approval has taken so long! The labeling division of FSIS has recently approved a pork-surimi product label and we believe some products are currently on the market on the West Coast.

Many different product applications of the minced fish obtained from deboning have been explored (e.g., Regenstein 1986). Many of these can be duplicated by running an inexpensive previously frozen fillet through a food processor to create mince at home (Regenstein and Regenstein 1982). All of this work is based on the premise that fish mince must appear in the marketplace completely on its own. If it is merely mixed with red meat or poultry, consumers will view the fish as a dilution of the product for economic reasons. Additionally, combining fish with meat renders the product unacceptable to certain vegetarian and kosher consumers.

"Fish farce," that is, the trim wastes and sawdust associated with fish processing, is being used in experiments to make traditional, fermented sausages in Germany. The 8-ounce size of Bilal sausage by Superior Food Company was sold at retail in the $1.79 to $1.89 range for eight 1-ounce sausage links. The product seemed to be a great success in the Washington, DC, area, but disappeared from the market after a short time. Apparently the fish company in Uruguay making the product was unable to procure sufficient fish to continue the project. Pacific King Foods has introduced a Wienk sausage in three flavors: shrimp, salmon, and vegetable using a natural casing; and Terra Nova Fisheries is planning a cod surimi sausage. Tuna ham and sausage are being produced in Costa Rica, and Norway has recently shown a herring-based sausage at trade shows. Israel has a tilapia-based hot dog. These are all very rubbery products, more like a surimi crab-leg (see below) than a true hot dog as usually prepared in the United States.

Deboning waste products can be used in fish meal plants. Although the removal of most of the flesh lowers the nutrient value of the resulting meal, the material still contains a large amount of potential flavor and could, for example, be boiled to make a fish broth. Use of vegetables such as celery, onion, and carrots not only adds to the broth's flavor but also helps mask some of the bitter flavors that sometimes appear in fish broths. The tuna industry uses the same trick when it uses "vegetable broth" in its products.

A variation of the mince process involves the deboning of menhaden mince after cooking to form a fish puree.

Many researchers and members of the fish industry have encountered the use of mince in the manufacture of surimi. Surimi deserves its own section in this chapter.

SURIMI

In Japan surimi is defined as a stable intermediate in the process of producing "kamaboko" products for the Japanese market. For centuries, a series of different Japanese products have been made from fresh fish minces. These foods are generically referred to as kamaboko—which is actually the name of the most popular of these numerous products. Although historically these products were always made from fresh fish, surimi as we know it today was developed in the early 1950s as a way of making a stable, frozen intermediate ingredient that could be stored while awaiting production into finished products. Surimi is considered by many to represent a great revolution in the fish marketplace. Whether they have been aware of it or not, American consumers have been buying and eating surimi in a variety of restaurant and fast-food products that have names like "imitation crab legs" or merely "seafood salad." The advantages and disadvantages of using surimi are currently under great scrutiny and are the focus of great debate (Lanier and Regenstein 1986).

Surimi Production. Under normal circumstances, mince prepared from whole fish generally contains blood from the frame as well as black belly membranes and some nerve ("white worms") tissue when it comes out of the deboner. It may even contain an occasional bone or scale which can be removed by a screening process originally designed as part of the Bibun deboning equipment, sometimes referred to as the second stage.

To remove any unwanted color or fat, the mince can be washed a number of times in soft water; the presence of calcium or magnesium is detrimental. For surimi from dark-fleshed fish, the "whiter" meat is used to obtain a lighter product. A pH between 6.8 and 6.9 is ideal, offering a good compromise between the efficiency of washing and the gel strength needs. Removing the soluble proteins and the small-molecular-weight compounds requires potable fresh water, not salt water, and creates a high BOD (biological oxygen demand) waste stream. It also means that the nutritional value of the material is decreased by the loss of protein, vitamins, minerals, and other water-soluble materials. At sea, with very fresh materials, just one wash is used. Onshore, with slightly older materials, two or three washes are needed. Current research suggests that countercurrent washing flows might serve as an alternate method. Freshwater fish and saltwater fish also respond differently.

The material must be dewatered following each washing. The final dewatering works best if the water's temperature is above 10°C (50°F). Various screening and centrifuging methods are used for this purpose, after which the remaining material is mixed with various cryoprotectant agents that generally lower the gel strength. With all of the washing of the surimi and the removal of the supposed substrate TMAO from the surimi, it is difficult to understand why it still requires all of the cryoprotectants. Presumably the changes that take place during frozen storage are different, possibly more related to the normal denaturation of all frozen muscle. Further work with these materials might help elucidate what is happening.

Like regular fish, surimi does better if the temperature of frozen storage is lower. For maximum stability, −35°C (−31°F) is recommended; −20°C (−4°F) is not as good, and higher frozen storage temperatures are even worse.

Components of Surimi.   The most common surimi formula is 92% washed mince, 4% sugar and 4% sorbitol, which creates a fairly sweet material that can be stored for up to a year in cold storage with minimal loss of gelation properties. The sorbitol helps to reduce the amount of sugar needed, thereby controlling browning and sweetness, but adding a lingering sweetness associated with sorbitol. Polyphosphates (0.2%) are often included at this stage. Whether the polyphosphates have an actual "cryoprotectant" effect or not is currently being debated. Polydextrose, a nonnutritive bulking agent, lowers the pH of the mince/surimi. A neutralized version is available that seems to be a more effective cryoprotectant. It has been suggested that the salt and water activity of surimi may help inhibit the growth of *Clostridium botulinum* under abuse conditions.

In the Japanese marketplace, the best surimi is believed to be that made with the freshest materials; manufacturers consistently pay a premium for the various at-sea grades of surimi. Until recently there was only one lower-grade available for purchase from among all shore-based surimi plants. Foreign markets have started to exert pressure to establish additional shore-processed surimi grades. The United States, in particular, has been developing a shore-side industry and exports surimi to Japan now that Japan has lost many of its traditional fishing grounds as other countries have extended their fishing boundaries to 200 miles. The Alaska Fisheries Development Foundation has helped to start an American surimi industry by providing funds through the NMFS for purchasing and operating the first surimi plant in Alaska. Unfortunately, after the equipment was first installed in a plant in Dutch Harbor, the company folded and the equipment had to be moved and reinstalled in a plant in Kodiak, Alaska, where it is now functioning well.

This plant (Alaska Pacific Fisheries) can now produce a very high-grade surimi but it is still purchased by Japan as "shore-based" product. Because the fish used are much fresher than those available to the Japanese shore-side processors, the Alaska product is of a higher quality: it is one day old versus four to five days for Japanese shoreside facilities. However, the American product is made from fillets coming off a Baader 182 filleter. These fish fillets would sell for $0.65/lb as is. After they have been minced and processed into surimi, their wholesale price becomes very variable, ranging from $0.45 to $0.95 per pound. The process also generates a high protein–vitamin–mineral waste—making the entire undertaking economically questionable.

Currently many more plants have been built in the United States both to produce surimi and to further process it into products for the domestic market. Most of the surimi is still being exported to Japan. The shore-based surimi manufacturing companies in Alaska are still partially or wholly Japanese owned. Additional processing of surimi is taking place at sea on the factory trawlers. Because of various laws concerning such ships, these are American vessels with American crews.

The major species used to make Japanese surimi has been Alaskan pollock, a fish which is widely distributed throughout the North Pacific Ocean. Over sixty other species are used for kamaboko, but apparently pollock is the main one routinely used for surimi, although some commercial hoki and blue whiting surimi are being used. When it is available, croaker from the Gulf of Mexico has been used to produce frozen surimi. A project in Virginia to make surimi from menhaden, a dark-fleshed fish, was unsuccessful. The low yield of washed mince, often less than 10%, made the process uneconomic.

Until recently, Alaskan pollock was a very underutilized species in the United States, possibly because it is notoriously infested with seal worms. The U.S. pollock resource probably represents the largest single stock, or biomass, of fish anywhere in the world. However, because of the worms, which will vary depending on the location of the catch, its use in fish blocks requires very careful candling; mincing the pollock merely disregards a problem that will not just go away.

Editorial note: Many of the proposed surimi products might eventually be appropriate for the kosher market, for example, crab legs with no shellfish. However, the presence of the worms would limit this possibility. At present, the more serious limitation is the fact that most, if not all, of the flavors being used in surimi products are derived from (nonkosher) shellfish by-products. Recently a series of products have appeared in this market and are labeled as "Not-really Crab" and the like. In addition a generic "white fish," for use either in preprepared salads or for making salads in the home has also appeared on the market, both in kosher and nonkosher forms. (See Chapter 16 for more details on kosher and nonkosher fish.)

Current research includes efforts to develop methods of preparing surimi from other species, a few of which as indicated above have been commercialized, including blue whiting from the Faroe Islands and the United Kingdom, silver smelt from Iceland, pelagic fish from Peru, hoki from New Zealand, red hake and whiting from North Atlantic, menhaden from the United States Atlantic Ocean and Gulf of Mexico, and mackerel and croaker from the shrimp by-catch in the Gulf of Mexico. An effort to produce a squid surimi has been stymied by the activity of the squid's proteolytic enzymes.

This impressive variety of species highlights the fact that the fat content of the fish used does not seem to be a limiting parameter in our ability to make surimi, probably because most of the fat is washed out during processing. A few years ago, the defatting would have been considered a health benefit. However, recent information about the advantageous role of omega-3 fatty acids (Chapter 14) has changed this perception.

Surimi Products. All of the numerous Japanese surimi products have a strong rubbery texture when cooked. This gelling (setting) phenomenon is brought out by the washing and subsequent addition of cryoprotectants and is the key to the success of surimi as an intermediate for kamaboko and other products. This functional process is currently the focus of much research.

For the most part, Japanese surimi products are considered too rubbery (and sweet) for the traditional American palate, although as Americans embrace many different ethnic foods, a market for the traditional Japanese product may be developed and the product may be marketed beyond the traditional Oriental communities. Preliminary research in the United Kingdom sought a technology—including a partial washing of fish mince—to yield a more stable material; washing out the small-molecular-weight compounds like TMAO might create a more stable material without a rubbery texture. The raw material can be used for products such as fish nuggets.

We are now seeing Japanese products in the United States such as "kamaboko," simply an extruded loaf of surimi which is then steamed. Part of the product is sometimes dyed red and coextruded to offer a little more visual interest, for example, a jelly-roll type of appearance. The product can be broiled ("chikuwa") or fried ("satsuma-age)." Ingredients such as vegetables, shrimp, or pork are often added to the final product that is marketed as "Tempura" in the United States. Interestingly, for the Korean market in the United States, the manufacturer of these traditional Japanese products must add some regular fish mince to give the product more flavor.

However, the major thrust of surimi development in the United States and other Western countries has been in the development of imitation seafood products, that is, seafood analogs. Use of the label "imitation" is a legal requirement in the United States because these products are not equivalent nutritionally to the seafood products they are replacing. They are higher in salt and lacking in some vitamins and minerals that are washed out. The most successful product to date has been the imitation crab-leg. One of the methods of its production is to extrude and cook thin sheets of surimi. These sheets are sliced, rolled, covered in film, recooked, and then cut and packaged. The time/temperature relationships of the original cooking process are critical in this process.

Other imitation crab-legs are produced by direct extrusion. These are, of course, cheaper to make. Some people feel that some of these products are also of low quality and that while these products may initially compete in the marketplace for the attention of the price-conscious buyer, whether this is the final consumer or an institutional purchaser, they will eventually lead to consumer dissatisfaction.

Legal Issues.   Other surimi products are sometimes used in food service or institutional environments and are often called "seafood." There is some concern that this is misleading and the FDA may be altering this terminology in the near future. The FDA has recently rejected a petition to remove the "imitation" labeling on surimi with 15% less protein than the seafood it resembles. The National Fisheries Institute's (NFI, the trade association for seafood processors) Surimi Committee is trying to get the term "Surimi Seafood" legalized.

The FDA has also issued a tentative standard for plant proteins that are made to resemble or substitute for seafoods. Presumably surimi would have to meet a similar standard.

Note that the FDA guidelines only refer to those final products that are analogs of other seafood products. Others have no special requirements. These, however, do not seem to be having a great deal of market success.

The State of Maine has recently passed a law which requires special labeling for all products made from surimi. The National Fisheries Institute's Surimi Committee is looking at approaches for alternate naming schemes. The New England Fisheries Development Association is looking at consumers' perceptions of the product.

Marketing Issues.   The timing of the introduction of imitation crab-legs was excellent. Overfishing of real crab populations had decimated the natural crab fisheries; king, snow, and dungeness crabs were all down in availability. The relatively high price of the natural crab products in cold storage does not compete well with the imitation crab-legs. (Note: At least one surimi product on the market claims to be made with real king crab. Given the high price of this product, it is likely that a species of crab other than that which most Americans would call "king crab" is being used. Presumably it is still a "centrolla" crab. Given the industry's lack of standardized nomenclature and the current inability of the technology to specifically identify fish and shellfish species, the product is still allowed in the marketplace.)

Current efforts have been directed towards developing imitation shrimp, scallop, and lobster products. Most of the flavorings used are still derived from seafood extracts; thus, people who cannot eat seafood because of allergies or religious prohibitions must still avoid these products. American flavor houses, that is, companies that specialize in producing flavors for the food processing industry, are working hard to find other sources for surimi product flavors that will find favor in the marketplace.

Most American surimi is found in "seafood" salads. If this trend continues, then the development of other forms may increase consumer interest, but will not necessarily lead to more sales. However, sales are in fact sometimes increased by this heightened consumer interest, but cannot be used to make very long-term predictions about the product's success. In the classical product life-cycle, a new product starts with low sales and rises to some upper limit; this is followed by a fall-off as the product finds its long-term sales level, which also gradually decreases over time (see Figure 9.4).

Until recently, most surimi manufacture and processing occurred in Japan and the resulting products were imported into the United States, both raw materials and ready-for-market imitation seafood products. This is now no longer true. A small plant in Bayou LaBatre, Alabama (Nichibei Fisheries), was the first plant to be equipped to make surimi but had to close down due to an inconsistent supply of croaker from the shrimp by-catch. Nevertheless, many of the early experiments with other sources of surimi were carried out in this plant, with the NMFS funding much of the equipment. Some of the Japanese companies have established further manufacturing facilities in the United States—specifically Bibun in North Carolina and Washington state. The Alaska Pacific plant in Alaska is actually owned principally by Japanese interests, although its equipment was also funded by the NMFS. A second Japanese-owned plant is operating in Dutch Harbor, Alaska. Additional plants and at-sea processing vessels are currently capable of producing surimi. In some cases, the facilities plan to alternate between surimi and pollock fillets, depending on the market price of each item.

The original American-owned company to produce surimi-based products was the Japanese American Company, producing imitation seafood, and its sister company, Yamasa Enterprises, producing Japanese and other Oriental products, both in Los Angeles. Frank Kawana, a Japanese-American, is the owner of Yamasa Enterprises and has been very active in promoting the Americanization of surimi production. Referred to by some observers as

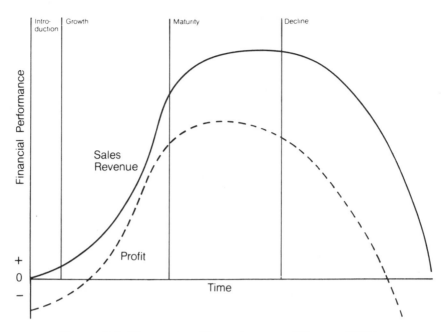

Figure 9.4 The product life-cycle curve. (*Chaston 1983*)

"Mr. Surimi," Mr. Kawana had also helped establish the Bayou LaBatre plant. His company was recently bought out by International Multifoods in Minnesota. Another Minnesota company, Kemp Foods, has been bought out by Oscar Mayer (Philip Morris: General Foods/Kraft) and marks the entry of a large corporate conglomerate into the surimi-products business. A number of other companies are currently making surimi products, but some observers believe that there are too many companies in the market and that some consolidation will have to occur in the near future.

In summary, for every advantage of surimi, the authors find at least one disadvantage. It is true that surimi can use previously underutilized species; but it does so at great processing expense. One account suggests that it takes—and costs—20% more in labor to produce surimi than it takes to inspect and pack fillets. Surimi can sometimes substitute for unavailable popular species; but on other occasions it can hurt the "real" seafood market. Surimi can be a boon to consumers who are allergic to seafood, which allergy is more common than sensitivities to other fish; but surimi is most successful in the United States only with the addition of numerous flavoring additives derived from shellfish. There are also general concerns about legal and ethical labeling and marketing. In all, use of surimi is not a simple panacea to "whatever ails" the fishing industry. Two excellent references on surimi are Lee (1984) and Suzuki (1981). (Note that the present authors do not recommend Chapter 1 of the latter text.)

A TECHNICAL NOTE ON SURIMI: GELATION

Gelation is the process whereby a material sets up an extensive network that can trap within it a large quantity of water. To determine the amount of gelation that takes place in cooked surimi, a disk of kamaboko of specified dimensions is prepared in a sausage casing, then cut and folded in half once and then a second time. The ability to survive both folds indicates the highest grade of surimi; breakage at various points defines intermediate grades. The property being measured is referred to as the "Ashi" of the kamaboko.

The gelation process for fish can occur under three different temperature conditions:

1. A cold setting occurs if the product is left in the cold room for a while. A similar setting takes place with other meat (flesh) products and is the reason that most further processed meats are mixed and made in one operation.
2. A stronger final gel, at about 90°C (194°F), is obtained if the product is first heated to 40°C (104°F) for about 30 minutes. This is referred to as "suwari." There has been some argument as to whether this process is unique to fish. It occurs at about 3–5°C (37–41°F) above a very obvious transition observed in the differential scanning calorimetry (DSC) traces of the fish being heated. The gelation process does not occur at this temperature in either red meats or poultry. These experiments should be repeated at a slightly higher temperature, that is, at 3–5°C over the same transition in the meats, where it occurs at slightly higher temperatures.
3. However, if the product is heated to 60–70°C (140–158°F) for about 30 minutes, the proteolytic enzymes in many of the fish species cause a weakening of the final gel structure either at that temperature or at subsequent heating to 90°C (194°F). The 60–70°C range is normally used in red meats and poultry for hot dog and bologna-like products and suggests a real limitation to any potential use of fish meat in these products.

Gelation research has led to the development of additional tests for measuring this property in surimi. One system being proposed measures true stress–strain information about the products rather than using empirical tests such as those we have described earlier. Unfortunately, the tests require special, customized equipment.

THE NORWEGIAN FLOTATION PROCESS

The processes described above generally use gutted fish, and are therefore not economical for use with small fish, especially the pelagics such as herring and capelin. Researchers at the Norwegian fish laboratory in Tromso have attempted to develop an alternative approach for the smaller fish. They take the whole fish, for example, herring, and chop it into slices (steaks) approximately 1 inch wide. The fish pieces are then placed in a water/acetic acid pool. Very careful control of differential flotation conditions such as salt concentration and pH permits the fractionation of the fish: the fish pieces sink and the fat and viscera tend to float; a separate flesh component can be obtained and the fat is removed. The resulting fish food is used in "mince" applications. In fact, it may still have to be run through a deboner to remove any attached bones. Current processing is done at 30°C (86°F) with a pH of 4, conditions which, unfortunately, destroy some of the functional properties of the fish mince. Work is in progress to get the temperature below 30°C and the pH above 5.5.

Experiments with proteolytic enzymes are also being done. Enzymatic deskinning of herring can be accomplished by giving whole herring a short wash in dilute acid, thereby denaturing the surface collagen and killing bacteria. An acid protease from the stomach of cod is then used to digest the skin. Squid can also be cleaned this way, either with naturally available enzymes or with a trace enzyme found in papaya latex; the latter even works at 0°C (32°F). The intestines of fish contain an enzyme which removes the roe from the ovary tissues.

COMPOSITE FILLETS

The manufacture of composite fillets is another way of using both small fish and broken pieces of fish. Various forming machines mold these materials into specific portion control units of any size or shape. Different machines tear the flesh to varying degrees, and this tearing affects the fish's final texture. The amount of fish fiber orientation desired in a final product may also affect the choice of equipment. One model uses a hydraulic ram; another uses an auger-feed; other types are also available. Panel surveys indicate that large piece size, and true orientation of fibers would improve the product, but no equipment currently exists to effect these changes.

The products are generally pressure-formed so that they stick together. The amount of bind between pieces can be controlled by the addition of salt and/or polyphosphates. The product is frozen as soon as possible after its formation. The ability to generate portions of uniform size and shape is a real advantage of this process.

REFERENCES

Chaston, I. 1983. *Marketing in Fisheries and Aquaculture*. Farnham, Surrey, England: Fishing News Books Ltd.

Lanier, T. C., and J. M. Regenstein. 1986. Surimi: Boon or boondoggle? Two food scientists offer differing views. The key to surimi is functionality (Lanier); Why not use minced fish instead (Regenstein). *Seafood Leader* (Winter) 152, 154–155, 157–158, 160–162.

Lanier, T. C., and F. B. Thomas. 1978. *Minced Fish: Its Production and Use*. Raleigh, NC: University of North Carolina Sea Grant College Publication.

Lee, Chong. 1984. Surimi process technology. *Food Technol.* **38**(11): 69–80.

Regenstein, J. M. 1986. The potential for minced fish. *Food Technol.* **40**(3): 101–106.

Regenstein, Joe M., and Carrie E. Regenstein. 1982. *Choose Your Title: Kosher Minced Fish Cooking; Fish Cooking with a Food Processor; International Fish Recipes*. Albany, NY: New York Sea Grant Bulletin.

Suzuki, T. 1981. *Fish and Krill Protein Processing Technology*. Barking, Essex, England: Applied Science Publishers. (Note that the present authors do not recommend Chapter 1 of that text.)

10

Industrial Fish Processing and Related Processes

The usual waste products generated by the fish processing industry are fish meal and fish oil. Because these products have significant economic value, fish meal and oil may also be produced deliberately from the oilier pelagic species. Referred to as "industrial" fish, these species can be caught economically in large volumes, that is, they are "schooling" fish, and are generally not handled in the traditional labor-intensive way on board.

FISH MEAL

Common industrial fish species include menhaden, herring, capelin, pilchard, and anchovy. Because fish destined for fish meal are generally not well iced at sea, preservatives such as formaldehyde may be used to prevent spoilage. The formaldehyde is usually used in a solution called Formalin. Consumer concern about the safety of this compound is related to high levels of vapor in the air and its subsequent interaction with the lungs. However, in protein foods, the formaldehyde has already reacted with the protein and other biologically reactive materials and is not available as chemical formaldehyde. Recent work suggests that potassium sorbate might also be an effective preservative for fish destined for the meal plant. Other research shows that propionic acid may yield reduced drip loss and a lower protein content in the resulting drip when used in storage of industrial fish prior to meal production.

Production and Storage. Exactly what happens in the classical fish meal process, referred to as the "wet fish pressing method"? First, the fish is cooked to coagulate the proteins so that they will not emulsify the oil. Fish that are too fresh hold water too well and are not well suited for the fish meal process. It is difficult to age fish properly for meal processing without allowing them to spoil and yield a much lower-quality fish meal.

Fatty fish may have their fat located in adipose cells, surrounded by a collagen layer, which melts at a lower temperature than this cooking temperature. It has recently been shown that some fatty fish, for example, capelin, have the bulk of their fat below the skin and in the belly flaps; others for example, mackerel and herring, have their fat cells located throughout the muscle tissue.

In the second step, the fish are pressed, yielding two products. The liquid that comes through is called *presswater*. The remaining pressed material is called *presscake*; it is dried to become fish meal. Both direct and indirect drying methods are used. The latter seem to give a higher-quality meal. Currently, lower temperatures of cooking are being used, again to yield a higher-quality meal.

The presswater is desludged, usually by straining, to remove any solid particles. These can be added back to the presscake. It is then run through a centrifuge to separate the oil from the water; the oil is refined and sold as fish oil. The aqueous phase may sometimes be discarded, but it contains many of the essential vitamins and minerals of the fish flesh. The product is called *stickwater* at this point. In the processing of stickwater, it is important to keep the temperature below 150°C (302°F) so as not to destroy vitamins and heat-sensitive amino acids. Three evaporators in series with progressively higher degrees of vacuum are generally used to concentrate the stickwater. This concentrate can be sold as "fish solubles" or spray-dried and added back to the fish meal to improve the meal's nutrient content for feeding monogastric animals. A typical flow-sheet is shown in Figure 10.1.

With low-fat fish, it may be possible to prepare the meal by simply cooking the fish and drying the meal. An unusual variation of the fish meal process is being explored in Japan. The meal is fried first in the hope that the frying might seal-in the vitamins, minerals, and fat while simply volatilizing the water. The meal and oil are separated by pressing after the frying.

During the storage of fish meal, the oxidation reaction of the remaining fat produces heat. If the material is not properly cooled, the fish meal can undergo spontaneous combustion. Thus, fish meal should generally be less than 10% fat/oil; an antioxidant such as ethoxyquin is often added at the 400–700 ppm level. TMAO and TMA may be part of the natural antioxidant system of fish meal and, by extrapolation, possibly of fish itself.

Uses of Fish Meal.   Because of its amino acid composition, fish meal can complement other feed sources very effectively. For example, fish meal, deficient in isoleucine and the sulfur amino acids, combines well with sunflower meal, the presscake following extraction of sunflower oil.

Not all uses of fish meal have been successful. For example, when some fish meals have been used in large quantities in poultry diets, there have been reports of gizzard erosion. This is especially true with a higher-temperature treatment of a fish meal with large amounts of lysine, histidine, or histamine. The compound in fish meal causing gizzard erosion has been

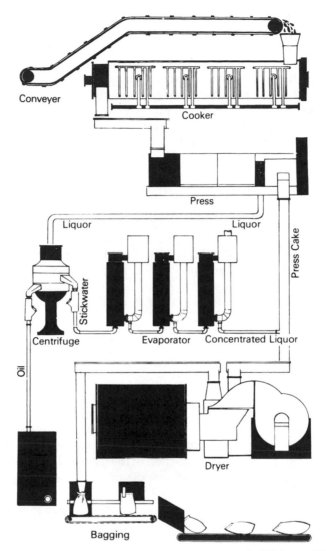

Figure 10.1 A typical fish meal process. (*Aitken et al. 1982, p. 153; Crown copyright*)

called gizzerosine, believed to be created by the reaction of histidine with the protein in the meal. A diet of 70% fish meal and 1% lysine for 1 week for 4–14-day-old chicks is recommended as an assay for gizzerosine. Although the extent of erosion may vary according to the species of fish being used to make the meal, heating above 130°C (266°F) seems to increase erosion in most cases.

Fish meal's high content of methionine and lysine cannot always compete with less expensive synthetic equivalents. However, research on young turkeys has continued to support the claim that fish meal may have beneficial, though unidentified, growth factors, that is, nutrients that have not yet been discovered but which make various animals grow more rapidly with fish meal than with other feeds, particularly plant-based diets.

For many years fish meal was considered particularly desirable by feeders of monogastric animals because of "unidentified growth factors." On the other hand, current nutritional information indicates that this advantage of fish meal no longer exists for most commercial animal species. For example, trout—considered by many to be essentially carnivorous—can now be raised successfully, that is, with the same growth rates, feed efficiency, and edibility characteristics, on a plant-based diet.

When used as a feed ingredient, fish meal must generally be kept below 10% of the final diet of the monogastric animal in order to avoid taints. That is, the use of fish meal often leads to a fishy flavor developing in the flesh of the animal. It is also important to avoid salmonella contamination of the meal. The better the quality of the fish meal, the less chance of taint. Some of the early unsuccessful efforts to use fish meal were made with fish meals that were not properly stabilized, that is, they did not contain ethoxyquin or any other antioxidant.

Recent feeding studies at home and abroad suggest that there may be a place for fish meal in ruminant diets. Small amounts, that is, 1–3% of the diet, may improve both feed efficiency and the lean/fat ratio. One estimate by researchers in Cornell University's Animal Science Department is that the "least cost value" of fish meal could be over $800 as compared to the general selling price range of $250 to $450 per ton. However, although a small amount is very beneficial, if the level becomes too high problems will be encountered. It is also important to time the introduction of the fish meal diets. They probably should not be introduced during the middle of a lactation period.

"White" Fish Meal.   "White fish meals" are those made from by-products of filleting traditional white fleshed fish; they tend to be low in zinc, or possibly zinc availability, as compared to regular fish meal. The supply is also much more limited at this time.

For effective use as ruminant feeds, white fish meals may have to be prepared without the addition of fish solubles. Ruminant nutritionists believe that the most important factor governing the ability of ruminants to use feed proteins of high quality is the ability of these proteins to bypass the rumen. Therefore, a certain amount of indigestibility is desirable. The protein should be able to pass through the rumen without being digested, but should then be digestible in the true stomach. Solubles by their very nature do not fall into this category and are therefore probably not desirable in ruminant feeds.

FISH OIL

The other major by-product of the rendering process is fish oil. The highest quality fish oils are obtained from the freshest quality fish and have almost no free fatty acids. In some countries, if the process is approved by the appropriate government agency and the oil is of high enough quality, it may be hydrogenated and used directly as part of the human "edible oil" supply, often as an ingredient for foods like margarine.

For oil of human consumption quality, it is also important that the starting fish be of high quality because many of the bacterial breakdown products formed during fish spoilage—for example, the various sulfur compounds—inhibit the catalysts used for the hydrogenation reaction. Current research (NMFS, Seattle) includes experiments with the molecular distillation of ethyl esters of fish oils as well as the storage of these oils under nitrogen in pressurized cans.

Production.   Fish oils from fish skin can be made by boiling for 10 minutes in 13% hydrochloric acid, washing the oil three times, and filtering at 15–16°C (59–61°F). Fish oil from fish liver can be made from mince that is heated to 30–40°C (86–104°F) for 30 minutes and adjusted to pH 9–10 with sodium hydroxide at 85–90°C (185–194°F). It is then washed three times, stored at 0°C (32°F), and strained. According to a recent report in *Fishing News International*, cod liver oil is enjoying renewed popularity in the health food market.

Fish oils must be refined to remove the free fatty acids and other impurities before they are hydrogenated. The free fatty acids are measured by titration and arbitrarily expressed in terms of oleic acid (molecular weight 282), that is, as if each free fatty acid were an oleic acid molecule. The oil is washed initially with a concentrated (80–85%) aqueous solution of phosphoric acid. This degumming process removes phosphatides. A 4-molar sodium hydroxide solution is then added in about 20% excess of the free fatty acids to produce sodium salts that settle out and can be removed. A potential substitute is under investigation in South Africa: ethanolamine is a natural component of phosphatidylethanolamine. It forms an ethanolamine salt (soap) without forming an emulsion. Its economic potential looks promising since the ethanolamine can now be easily produced from ethylene oxide and ammonia. The liquid can be added directly without creating an aqueous phase. However, it may have to be followed by a sodium hydroxide step. Glycerol was also tried, but the results were not very encouraging.

Current Research.   At present the use of fish oils for human consumption in the United States is undergoing regulatory review. The feeding trials on animals required by the FDA for such approval have been done by the

National Fish Meal and Oil Association (NFMOA) in cooperation with a number of other agencies. The NFMOA has submitted a petition for GRAS ("generally recognized as safe") status for menhaden oil and partially hydrogenated menhaden oil. At the time of this writing, approval of a hydrogenated oil, which loses some of the most beneficial aspects of fish oil, has been approved by the FDA. Additional studies on the use of the unhydrogenated oil were requested and no final ruling has been announced. The potential for using fish oils in human food products improves as we gain more knowledge about the benefits of the oil's omega-3 fatty acids.

MARKETING ISSUES FOR FISH MEAL AND FISH OIL

Fish used for meal and oil production may represent as much as 35–40% of the total tonnage of fish caught worldwide. The price of fish meal is generally about 1.6–1.8 times the price of soybean meal. This is just a little too high for fish meal to normally appear in animal diets formulated by a "least cost" formula, that is, those diets formulated by modern computer techniques that match a set of diet specifications—a list of nutrients needed in the feed—with a set of analyses for these nutrients in the available ingredients.

Prices for both fish meal and fish oil tend to fluctuate. Fish oil in northern Europe is down about 40%, probably because of the increased availability and lower price of palm oil. Fish oil prices are generally lower than those of other oils because of the hydrogenation requirements. Fish meal, for example, increased in price following the scarcity of soybean and other oilseed proteins due to an extensive drought in the United States soybean-growing areas.

Some of the industrial fish can actually be used directly for human food, a decision that can have important political/social/economic ramifications. Recently, for example, certain Third World countries have attempted to get a better return on their industrial fish by moving more of the catch directly into the human food supply. However, much of this fish ends up in meal anyway because of fish quality problems and/or the greater availability of meal plants than of freezing/canning plants.

Although menhaden, the major industrial fish in the United States, is generally not considered to be a food fish in this country, recent research suggests that it could be used as a source of fish for surimi production. Another possible human food use would be a cooked, "flaked" product in a can. We have also had some success using the minced fish in pattie products, the key being the choice of the proper antioxidant. We use ascorbic acid or erythorbic acid rather than one of the traditional oil-soluble antioxidants that one would have presumed to work best.

There are other uses of fish by-products for animals. Shrimp by-products are useful for flavor component extraction and for livestock feed, particularly applications that take advantage of the color the meal can impart, such as in farmed shrimp and salmon. Drying of the meal reduces the astaxanthin content. Another feed use of fish by-products could be the inclusion of lobster shells or crab meal in high-lactose (whey) diets to prevent digestive problems. Apparently these products may aid in the absorption of whey by young heifers. This allows for the use of whey in place of corn and soymeal for up to 45% of the diet, in spite of whey's high lactose content; however, research results on its benefits are not consistent or compelling at this time.

Sometimes, though, new and unusual uses of fish by-products become practical, for example, dogfish liver, high in squalene, has recently been found useful in the Japanese cosmetic industry.

Environmental Issues.   The production of fish meal and oil is often a smelly business, making the fish meal plant an unpopular neighbor. Environmentalists are justifiably concerned with the potentially serious waste disposal problems of filleting operations. In 1985, the major northeast fish meal plant located in Gloucester, Massachusetts, was closed: now that the community's economy is based on its summer resorts and tourism—rather than on fishing—the strong fish odors could no longer be tolerated. The SeaPro plant in Rockland, Maine, was the last plant in the United States northeast to produce fish meal and oil. Following a nonbinding referendum that voted 3 to 1 against continuing the meal plant, the company chose to close the plant. The environmental concerns are so great that industrial journals continue to publish many articles on the subject, including alternatives (see later) and ways of improving meal plants from an environmental point of view.

When the Gloucester plant closed, fish filleting wastes from plants throughout southern New England were dumped out at sea using the "gurry" barge until funds for this project were discontinued. The procedure was carried out with a temporary permit from the EPA (Environmental Protection Agency), and was subsidized by the Commonwealth of Massachusetts. The material had to be trucked to the barge; a charge of $20 a ton was required for barging charges. The method could not be applied to frozen product. Thankfully, the winters during the barging operation were not too cold. There is still a great deal of concern about fish waste in the northeast, but with the decrease in fish processing there due to declining fish stocks, most of the excess gurry has found markets as pet food and mink food.

In an effort to cut down production odors, the now closed SeaPro plant in Rockland, Maine, experimented with passing its vapor through a compost pile made of heather and peat. The bacteria in the compost are supposed to degrade any proteins and other materials in the vapor. Theoretically, the unit remains operable for 3 years. Results are not yet available.

Peat is a remarkable material. It has a tremendous water-retention capacity, holding about three times its weight in water. Because of its acidic nature, it is also a good absorber of ammoniacal odors. Its use in composting will be discussed later.

OTHER WASTE DISPOSAL PROCESSES

The major alternative for economic disposal of fish processing waste is the use of fish waste as a pet food, or as a feed for other carnivorous animals, particularly mink. The selling price is only four or five cents per pound of frozen and packaged feed. Most mink feeding occurs during the warmer months of the year. However, this disposal method is not available in the colder seasons when alternate methods such as land applications are also not available.

In spite of Pilgrim folklore, it has become clear that earning more than $0.05/lb on fish waste-as-fertilizer requires applying it to the land just prior to turning over the earth so that the fish waste is not kept on the surface where it can smell. Various processes for stabilizing fish waste, for example, hydrolysates and compost, have the additional benefit that the process can be carried out at all times of the year and the material can be saved, without smelling bad, until it is needed.

In many places, fish by-products are going to landfill because a special permit is required to dump fish waste at sea in the United States. The use of fish by-products on land can cause community problems if odors are allowed to develop. Yet in some cases, edible fish are landfilled. In at least one state, the Department of Natural Resources, choosing to have carp removed from their lakes to improve the recreational fisheries, paid 2 cents/lb for the whole carp and simply buried them in a landfill. The authors do not condone such actions.

Other technologies continue to be explored in the efforts to develop an odorless fish waste processing system. Newer fish meal plants have adopted modern odor-management practices. The real problem, however, is that most meal plants are quite old and the cost of upgrading them to current environmental standards is prohibitive given the limited value of the end-product. In locations like Kodiak, Alaska, the meal plant is only able to operate because of a continuous subsidy either from the government or from the packers themselves.

The Biotherm company has proposed a disposal process that is under scientific investigation in the New England region. The idea is to quickly use up the smelly elements of fish waste before the odors become too great.

Anaerobic digestion leads to the production of methane and a sludge: the methane is recycled into the process to produce the necessary electricity (heat); the sludge might eventually be useful as a raw material source of chemicals, but is currently only considered as a possible fertilizer. Unfortunately, the unit currently has a long cycle time, so that it must hold 20 days' worth of waste!

A second Biotherm project, recently reported, has been to make special fish hydrolysates under anaerobic conditions; these are quite smelly but show some unique insecticidal activity.

FISH HYDROLYSATES

Researchers continue to seek improved utilization of fishery by-products. One technique is to use proteolytic enzymes naturally present in the fish, or added enzymes, to hydrolyze the fish proteins; that is, proteolysis. A different process, acid hydrolysis, may also be used. In any case, these processes are delicate: differences in hydrolysates have been observed between wastes of fillet and whole fish, and between fresh and frozen product. The resulting product is used directly as an animal feed or as a fertilizer. In some processing plants, phosphoric acid is added to create a fish fertilizer for home use with potted plants, a product that draws a higher price than farm uses of fish waste. The phosphoric acid helps the digestion of the samples and also improves the final fertilizer value of the product by preventing spoilage and adding to its phosphorous content. Other acids are used, depending on the desired end-use.

One important technical consideration is whether the bones are screened out or not. As the bones dissolve, they require a great deal of acid to neutralize them.

If an enzyme hydrolysis is used, the product must subsequently be stabilized to prevent spoilage. Acidification is usually used. Sometimes the acid, especially lactic, can be generated by fermentation of carbohydrates. The final product may be drum- or spray-dried.

Self-digestion in the presence of acid is usually done with formic acid, and produces a product called *fish silage*. It is currently used to feed various monogastric animals such as poultry and pigs. The major problem with fish silage is that its production is only economic in the wet form and it is extremely expensive to move water very far. It is hoped that further work with ultrafiltration and/or reverse osmosis might make the product cost-effective.

In another attempt to use fish waste, silage is being incorporated into soft pellets for feeding aquacultured trout in Norway. The silage is mixed with a binder and an antioxidant; the pH of the product is between 4 and 4.5.

The Stinson Packing Company of Maine and the Triple "F" Company of Des Moines, Iowa, are developing the Insta-Pro Dry Extrusion process to produce a hydrolysate from partially defatted herring waste enzymatically digested by fermentation with molasses as the carbohydrate source. The product is being used as a pig weaner feed, that is, to replace sow's milk. They are also selling the stabilized herring oil removed by centrifugation from the hydrolysate.

These digestion techniques are similar to those used in the Orient to produce fish and bean pastes for human consumption. In fact, there are several "nonhuman" and "human" potentials for silage.

Shrimp waste can be silaged to yield a material with about 14–16% protein that forms a paste when mixed with other solid wastes, particularly carbohydrate waste—especially in the presence of certain organic acids. T. Chai at the University of Maryland showed that dogfish wastes can be ground and hydrolyzed by natural enzymes at about 50°C (122°F) at pH 4 to yield a dogfish paste upon cooking at 80°C (176°F). Biotec (Tromso, Norway) has developed a silage process, Ensiferm, that uses lactic acid bacteria (*Lactobacillus*?) and requires less acid. However, the process, which requires the waste to be pasteurized and fermentable sugar to be added, yields a very viscous silage. The presence of 0.1% potassium sorbate did not seem to help in a lactobacillus silage.

Even an oily silage can be used in the diets of pigs and poultry without taints if diets are properly prepared. Although both acid and fermented silages are variable, the latter require the addition of a carbohydrate source and a starter culture such as *Lactobacillus plantarum*. "Food" grade silage can be produced using propionic acid if the pH is kept slightly higher throughout the process.

It is hoped that further work with fish hydrolysates will continue to yield more materials usable as human food with flavors similar to those from hydrolyzed vegetable proteins or, at least, as high-quality nutrient supplements for animals.

FISH PROTEIN CONCENTRATE

Another way to upgrade use of industrial fish would be to produce a human-grade fish meal, that is, a fish protein concentrate (FPC). Common uses include incorporation into various baked goods. The process is, however, slightly different from that used for animal feed. An organic solvent is used to remove the oil, thereby preventing "fishy" flavors more effectively than would a simple pressing step. Hexane was the common solvent used for this purpose until it fell into disfavor because of concern about potential toxicity/carcinogenicity. Other approved solvents in the United States include isopropanol and 1,2-dichloroethane. Unfortunately, these solvents denature the protein almost as much as heat does, leaving a very nutritious but nonfunctional protein that cannot compete with soybean.

Use of ethanol has been slightly more successful because it permits the removal of fat and other objectionable materials without quite as much denaturation of the proteins. The ethanol is then redistilled and reused, for example, in the Japanese product called Marinbeef (Suzuki 1981). However, even this product gets mixed reviews when used as an extender in meat products like hamburger. In Peru, it costs more to produce Marinbeef than it does to produce steak; Peruvian Marinbeef is currently destined for cookies(!) in school lunches.

Ethanol is also used as part of the process to make omega-3-enriched fish oils. If 200 grams of urea dissolved in 800 ml of boiling ethanol is mixed with 80 grams of fatty acids, two layers form upon cooling. Urea crystals containing the saturated fatty acids can be easily removed by filtration. Laboratory results show a yield of 51% of the starting material containing 42% C20:5.

These fish protein concentrates are classified as FPC type A and generally aim for less than 0.5% fat in the finished product. The product has an interesting history. In the 1950s, FPC type A was touted as a possibly important protein source for underdeveloped countries, and the United States government approved the use of various "hakelike" species for this purpose. The definition of a "hakelike" species was essentially a species that was approved by the FDA for the manufacture of FPC; biological considerations were ignored. The industry's hope was to make FPC from smaller fish without having to gut them. At that time, the FDA did not approve the domestic use of this product made from uneviscerated fish; it could, however, be produced for export. This policy was considered unethical—and surely not politic—and helped destine the project to failure. Years later, the FDA approved the process for use at home as well as abroad but, by then, interest in the product had waned completely. The authors believe that the project would have failed anyway because FPC type A is too much like an expensive soybean meal. The reader is reminded that all fish meal almost always costs twice as much as soy meal.

Is there a fish protein concentrate that is more like traditional fish meal? FPC type B may contain up to 10% fat. It is similar to fish meal but meets higher quality/sanitation standards than the animal feed product. Although its fishy flavor is considered desirable in certain Third World countries, FPC type B has had very limited market success to date. This does not deter entrepreneurs from trying.

The FDA published a final regulation on fish protein isolate (FPI) in 1981 and then amended it to eliminate the microbiological standards in 1982. A petition to the United States Department of Agriculture to include FPI as a binder in sausages has been held up for three reasons: lack of data on effectiveness of FPI as a binder; safety of FPI containing meat products; and problems with the organoleptic acceptability of sausages containing FPI.

RINSE WATER RECOVERY

Much more work is needed in this area. The following is an example of what can be done. A recent process developed in Scandinavia deals with the problem of the high BOD (biological oxygen demand) in the wash waters of fatty fish being prepared for final products such as pickled herring. The wash water is prestrained to remove large particles; separators are then used to yield an oil/water phase which can be heated to separate out oil, water, and solids.

The solid phase, mostly protein, is currently underutilized, but may be usable as a feed. However, the oil is already economically viable and, following separation, the water phase can be more easily discarded. For example, with only 1.5% oil in the water, if 30 tons of water pass through the plant per hour, over 720 tons of oil are added per year into the municipal sewage system. Even with a 70% recovery, the amount of oil is sufficient to pay for the process.

REFERENCES

Aitken, A., I. M. Mackie, J. H. Merritt, and M. L. Windsor. 1982. *Fish Handling and Processing.* Edinburgh, Scotland: Her Majesty's Stationery Office.

Suzuki, T. 1981. *Fish and Krill Protein Processing Technology.* Barking, Essex, England, UK: Applied Science Publishers. (Note that the present authors do not recommend Chapter 1 of that text.)

11

Specialty Products and By-Products

ROE AND MILT

Roe, fish eggs, can be obtained from most species of fish. The most valuable roe comes from sturgeon and is called caviar. Although milt, fish sperm, can also be pan-fried and eaten, it is not a popular product in the United States.

Caviar is a processed food that usually starts with a species of sturgeon such as beluga, osetra or sevruga. Salt is added to the roe. (Malossol caviar uses only 3–5% salt and is considered a low-salt caviar.) After 30 minutes, the eggs are separated from the egg sack membrane by passing the roe sack over a screen. The rubbing breaks the sack and the eggs pass through the screen leaving the membrane behind. The product can be pasteurized and should then be stored at $-3--1°C$ (26–30°F). In some countries, but not the United States, it is legal to add up to 6% borax to sweeten the product. Whitefish roe is sometimes referred to as "golden caviar" although it is illegal to use the word *caviar* alone unless the product is derived from sturgeon. Lumpfish roe is generally preserved with sodium benzoate and tragacanth gum; it is naturally dark green but is generally dyed black. Red salmon roe is also available commercially and is an important export product to Japan.

Borax is the crystalline, slightly alkaline borate of sodium. Not approved as edible by the FDA, it is best known in the United States for its use in laundry detergents.

Considered a fishery by-product, most fish roe is discarded in the United States. Development of uses for roe could add as much as 10% to a fish's yield during the roe season. This should be particularly easy for flatfish, since these are not gutted at sea in the United States.

Although roe can simply be pan-fried or boiled and drained through cheesecloth, most consumers think in terms of using roe as caviar. It is possible that the high cost of caviar has discouraged American usage of roe. It is also possible that health-conscious Americans are put off by the high salt and cholesterol found in the final product.

Roe enjoys far more popularity in Japan, where the roe from almost any sea-faring animal is eaten. This includes crab roe, which looks like a sponge;

the roe from sea urchins, "uni"; and salmon roe, "sujiko," among others. Herring roe on kelp is considered a delicacy. This occurs naturally when herring spawn in a kelp bed. Harvesting of herring roe and herring roe on kelp has become a big industry on the United States west coast, particularly in Alaska.

The capelin roe fishery for export to Japan is important in the Scandinavian countries. Capelin roe is about 30% protein and has 90 calories per ounce. The production of minced fish by flotation (Chapter 9) was designed in Norway specifically to process the left-over male capelin.

A popular product in various European countries is a roe paste or pâté that is sometimes smoked and is sold in a can or as a tube-packed spread. Cod roe has been especially successful in this market. The fish eggs are removed from the skin (membrane) by washing in fresh water. Various roes are used to make taramasalata in Greece: the roe is combined with olive oil and bread to create a very elegant spread.

A company in East Germany claims to have developed a "caviar-like" product from animal protein, odor compounds, and food-coloring. It is added by drops into hot oil, yielding the "crisp, slightly sticky, and characteristic taste associated with Malossol caviar. A second group, this time in Israel, claims to have made a "sturgeon caviar" from all-kosher ingredients (sturgeon is not kosher).

OTHER INTERNAL ORGANS

A number of interesting uses have been found for a variety of internal fish organs. The search for further uses of these parts continues.

There has been an attempt to develop chitterlings from sunfish intestines in the southern United States. Chitterlings are a popular Southern dish, normally made with pork intestines.

The livers of many "low-fat" fish are well known for their oil and vitamin content. Cod fish livers can be canned in fish oils; this is an area appropriate for further product development work. After oil extraction, the remaining presscake could also be explored as a new raw material. Before undertaking any such project, the livers must be checked for contamination, as the liver tends to concentrate any undesirable materials. Also, liver oils are high in vitamins A and D and the potential for vitamin A toxicity needs to be considered.

Fresh fish bones can be ground to make an excellent bone meal as a calcium supplement. However, the fluoride, lead, and other trace mineral contents of bones must be monitored continuously to insure the safety of this product.

The bones from canned fish can be eaten, especially when mashed first (see Chapter 8). Although fish bones area healthy, many consumers object to them and remove the bones after purchasing the canned product. This bias is so strong that a few companies recently sought and received permission from the FDA to produce a boneless, skinless canned salmon product. Special permission was needed because the product does not meet the standard of identity for canned salmon. The product is currently being heavily discounted, suggesting that it has not been a success.

NONFOOD USES OF FISHERY BY-PRODUCTS

The scales of fish are sometimes used for nail polish pearl essence.

Dried fish scales can be used as a flocculating agent, that is, they can be used as part of a wastewater treatment system to precipitate the solid material, particularly protein materials. Chitosan, or crustacean shell, can also be used this way. The advantage of fish scales or chitosan for this purpose is that the recovered material may have potential as an animal feed or a plant fertilizer. Scales are cheaper than chitosan because there is no need for any chemical processing in preparing them as a flocculating agent. Although scales have been used as 10% of a poultry feed, this entire process requires more testing. The more traditional flocculating agents are generally inorganic compounds containing heavy metals that are toxic at the levels needed in feeds. Chitosan has also been reported to have specific weed-inhibiting properties.

The skins of some fish can be made into leather, then processed and dyed to make handbags and belts of surprising strength. A company in Italy is marketing attractive fish leather products in which the "scale pattern" adds to the design. At least one small company in Alaska has a catalogue of salmon leather items available for sale as does another in British Columbia. Hagfish (slime eels) skins are popular in Korea.

Shark is especially desirable for use as leather. Each hide is removed carefully with a curved knife, then fleshed, that is, has all remaining meat removed. After being washed with salt water and salted, the hides can be stacked on a slanted platform to permit the water to drain off. Resalting is repeated after 4 days; this schedule must continue throughout the duration of the hides' storage. Unsightly sour spots are related to decomposition; excessive exposure to the sun can burn the hides.

A factor has been isolated from fish skin that displays all of the characteristics of human interleukin-1, a hormone needed in the repair of torn muscle tissue. Further research is necessary to take advantage of this drug.

It is interesting to note the 1986 utilization figures on fish waste in New England: 30% for bait; 30% for mink food; 15% barged to sea; 8% for fish meal; 7% to landfill, and the remaining 3% spread on farmland. For use as baitfish, especially for lobster, the return ranges from about 10 cents/lb for redfish down to 3 cents/lb for herring. Pet food producers are taking fish waste at cost. On the West coast the waste is often hydrolyzed with enzymes, spray-dried and exported to the Orient for aquaculture feed, especially for eels.

Rennin Substitutes

Rennin, the cheese clotting enzyme, is normally obtained from the fourth stomach of milk-fed calves. Increased cheese production in the United States has outdistanced rennin's availability, leading the dairy industry to seek alternate sources for this type of enzyme activity.

Fishery by-products such as fish guts, seal stomachs, and clam bellies have yielded some potentially useful enzymes, but none of these has been commercially successful to date. Neither have any reports suggested that the alternate enzymes make a better cheese than rennin. However, some of these enzymes may find a use in tofu, soybean curd, when it is used as a cheese substitute or in other such plant protein systems.

In the long run, successful rennin substitutes for use in dairy cheese production would more likely be provided by biotechnology. In 1989, the United States FDA approved a chymosin, rennin, produced by biotechnology.

Another possibility is to use the enzymes of fish as processing aids. These enzymes have been called "biological knives" and can be used for processes such as skinning fish (or other materials) and removing the roe-sack membrane. A fish biotechnology company in Tromso, Norway, is currently trying to exploit this market.

Gelatin

Fish gelatin has a viscosity curve very similar to that of gelatin from higher animals. Its gelatin set temperature is much lower, about 8°C (46°F). It also has less proline/hydroxyproline. Currently it is used mainly in nonfood industrial applications. The conversion of collagen to gelatin occurs at about 30–35°C (86–95°F).

Chitin

Chitin, the major ingredient of the crustacean shell, is an acetylated amino-glucose with a beta $1 \rightarrow 4$ linkage (poly-$\beta(1,4)$-N-acetyl-D-glucosamine) and can be converted to chitosan by deacylation. The shell is collected, stored, and eventually ground to reduce the particle size. Dilute sodium hydroxide is used to remove any protein; then the remaining shell material is rewashed. Hydrochloric acid is used to demineralize the shell, that is, to remove minerals such as calcium carbonate. The resulting rewashed and dewatered material is now pure chitin. The chitin is then deacylated using hot, concentrated sodium hydroxide, washed and dewatered again, and then reduced in size. The product is called chitosan.

Researchers are seeking possible uses for this waste product. Chitosan is already used as an edible flocculating agent in waste-water management applications. It may also find a use as a biodegradable carrier of other biological compounds, particularly in medical applications.

Other possible uses of chitosan include the following.

- Animal feeds (already approved by the United States FDA). At the 0.5% level, chitin allegedly decreases the animals' food consumption and increases the animals' carcass weight. One patented process removes the soluble protein, which can be used as a feed, and also obtains marketable calcium chloride during the demineralization step. The antibodylike behavior exhibited by chitosan encourages this possibility since there is increasing pressure to decrease the use of synthetic antibiotics at sub-therapeutic doses in routine animal feeding.
- Protection of certain commercial crops from various microorganisms.
- Improving the functional properties of packaging materials such as paper-board cartons.
- Removing heavy metals from water.
- Clarifying beverages such as wine. The flocculation removes suspended metallic phosphorus compounds.
- Making hollow fibers that can be used for processes such as the separation of alcohol and water.

Shells

Shells can be returned to the ocean's floor as a base for growing various molluscs, mainly oysters. Shells can also be ground up for use in animal feed: oyster shells are often added as a calcium source in chicken feed, since laying hens require a large daily intake of calcium. Shells can even be broken up and used as driveway "paving" material—without damaging the tires of cars!

It should be noted that piles of shells from processing plants can sometimes be a problem, partly because it is very difficult to remove all the flesh that is stuck in and to each shell. Such remaining material attracts various animals to the shell pile, including ground animals like rats as well as birds, especially seagulls. This situation leads to potentially unsanitary conditions. Zall and Chou (1976) developed a method of using retort heat to remove some of the flesh adhering to shells. Unfortunately, this process is expensive with respect to the amount of meat that is recovered.

In some coastal countries of the world there is a thriving market in molluscan shells. They are sought by collectors and jewelry lovers. Still others are used in the button trade despite the modern substitution of plastic. In early America, the native American Indians used "wampum," a shell-derived material, as a form of currency.

REFERENCE

Zall, R. and I. J. Cho. 1978. Production of edible foods from surf clam wastes. *Trans. ASAE* **20**(6): 1170–1173.

12

The Chemical Biology
of Fishes

The title of this chapter is taken from the name of two outstanding books by Malcolm Love (1970, 1980) of the Torry Research Station in Aberdeen, Scotland. The books summarize current information about the animal biology of commercially important fish species. The focus is on the chemical biology of these species during the major "life cycle" events, including death and the stages of *rigor mortis*.

This chapter uses these books and other resources to explore a few topics. For a more complete review of the subject, we strongly recommend Dr. Love's books including his latest (Love, 1988).

SPAWNING CHANGES

The yearly cycle of a sexually mature fish includes changes in physiology and feed habits that can also lead to unusual hormonal and induced-enzyme changes. For example, the fat/water ratio changes in some fishes as their body mass becomes reproductive tissue; the fat content of menhaden goes from 2% to 25% and back again in the course of one season. In other fishes, the protein/water ratio changes just as dramatically, for example, in gadoids. A fish, as mentioned earlier, has been landed that was 95% water post-spawning. During starvation, cod use white muscle and liver lipids for energy; thereafter, they use proteins from both the red and white muscles. When they resume feeding, they overcompensate on some compounds. When re-fed squid, cod form more muscle glycogen (lower ultimate pH); with herring, they form liver lipids.

PARTS OF THE FISH

The skin of fish contains melanin pigments, the end-products of phenolic oxidation and polymerization. These pigments respond to changes in

175

relaxation and contraction of the pigment glands to give fish different appearances at different times. These changes may come to reflect the nature of the bottom the fish are living on (serving as a form of camouflage); for example, cod from the ground south of the Faroe Islands is distinguishable from cod from the north side by its different skin color.

Some fish also have internal flesh pigmentation beyond that due to hemoglobin/myoglobin. These are generally carotenoid pigments that are not synthesized *de novo* but are procured from the diet. These compounds in fish may break down due to factors such as light and lipoxygenase activity. The aquaculture production of salmon and similar species depends on the feeding of shrimp, crab, or similar feeds that already contain the necessary carotenoid pigments. With a diet high in crustaceans, it is possible to grow trout with a color more akin to that of salmon; some "salmon trout" have already appeared in the marketplace. Other efforts use an artificial carotenoid pigment (canthaxanthin) but it tends to yield too red a color; it must also be listed on the product's label. Artificial astaxanthin (the carotenoid in crustaceans) has recently been produced synthetically and has found a market in aquaculture diets, although its use is currently illegal in the United States. In practice the FDA office in the Boston area has been enforcing these regulations on salmon growers in Maine, but other FDA offices (West Coast and the import divisions) have generally ignored the problem.

The overall nonprotein nitrogen content of fish differs between the two main classes of fish. The teleosts have between 9 and 10% of their total nitrogen soluble in solutions such as trichloroacetic acid (TCA); the elasmobranch animals have between 33 and 38%.

The sensory organs of the fish are generally located along the lateral line. The major receptors consist of vibration or mechano-receptors. It is estimated that a fish can detect chemicals at 1 part per billion (ppb). Chartreuse is in the middle of the visible range and seems to be the most visible underwater color.

The fins of fish serve as stabilizers and brakes.

The muscles of fish are found in blocks, *myotomes*, that are attached to the connective tissue, the *myocommata*, which then connects to the backbone of the fish. The number of muscle blocks remains constant throughout the life of the fish.

POSTMORTEM EFFECTS

Given that the ultimate pH, that is, the lowest pH reached by an animal following slaughter, is due to the breakdown of glycogen to lactic acid during anaerobic metabolism, it can be postulated that the fish that are the least stressed while alive and at the time of kill will have the highest glycogen level at death and thus the lowest ultimate pH. These fish, if harvested with minimal stress, might be expected to have problems with gaping.

These facts may offer a solution to quality problems with inshore trap-caught fish in Newfoundland. These are caught in the middle of the summer when the fish are well fed; presumably, they are landed with a minimum of stress such that the low pH may be causing gaping. A suggested solution is to get a mechanical shark and put it into the traps with the live fish! The additional exercise should lead to a depletion of the glycogen and might yield a higher-quality fish.

COLLAGEN AND MUSCLE

Collagen is the major connective tissue of muscle and is the most ubiquitous protein in the animal kingdom. The collagen content of cod constitutes about 2% of the fish by weight. Figure 12.1 illustrates the pattern of connective tissue in fish; Figure 12.2 shows the variation in cell diameter of the myotomes of fish as a function of distance from head to tail.

One of the reasons that fish are not as tough to eat as other meats is that they have much less collagen than higher animals, but there are other reasons. The shrink temperature, that is, the temperature at which connective tissue undergoes a reversible elastic shrinkage leading to an organoleptic texture-toughening is lower for fish collagen. So is the melt temperature, that is, the temperature at which the connective tissue undergoes a nonreversible change from collagen to gelatin leading to organoleptic texture softening. Normal cooking of fish melts its collagen, eliminating the need to be concerned with cooking methods that tenderize the connective tissue, a concern that often exists with red meats. Fish species such as shark have as much as 10% of their body weight as connective tissue.

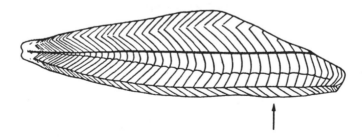

Figure 12.1 Appearance of the musculature of *Gadus morhua* after removal from the skeleton. The lines represent the connective tissue and the arrow indicates the region where the contractile cells are the longest. (*Love 1970*)

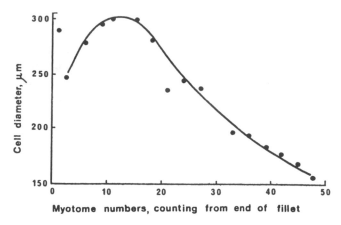

Figure 12.2 Average diameters of muscle cells from different parts of the musculature from a specimen of *Gadus morhua* 95 cm long. The myotomes are numbered from the point of severance of the head. Each point is the mean of 200 cells. (*Love 1970*)

In some species of fish, the amount of collagen and the amount of cross-linking of the collagen increases with starvation. Thus the toughening process in fish seems to be much more reversible than it is in higher animals, where the amount of cross-linking increases with age, thereby yielding older animals that are usually tougher than younger animals.

The collagen from sturgeon swim bladders is used as the source of isinglass. According to the song the "Surrey with the Fringe on Top" from Roger and Hammerstein's play *Oklahoma*, isinglass was used to make carriage windows in Oklahoma around the turn of the century. Today isinglass is mainly used as a clarification agent, particularly for beer.

The strength of the connective tissue is sensitive to pH. If the pH drops too low, the connective tissue breaks down. For example, this breakdown occurs below pH 6.4 in cod and leads to a phenomenon known as gaping. *Gaping* is the splitting open of the fish flesh, usually along the myocommata lines. Figure 12.3 illustrates gaped and ungaped fish.

The dark muscle along the lateral line constitutes the main swimming muscles of the fish. Because the dark muscle is more aerobic, the amount and often the darkness of dark muscle increases in those fish that do the most swimming. Not surprisingly, then, migratory fish are particularly high in dark muscle. This muscle is quite large in tuna, and is sometimes referred to as the "blood meat." It is not currently being used for human food because it is so dark and strong in taste.

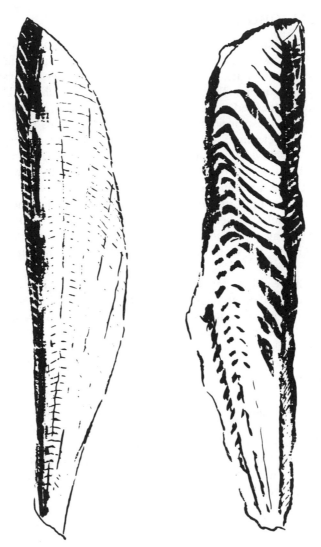

Figure 12.3 Ungaped fillet (left) and gaped fillet (right). (*Aitken et al. 1982, p. 6; Crown copyright*)

On the contrary, a fish like flounder moves slowly along the bottom of the ocean and has almost no dark muscle. Although the dark muscle is higher in myoglobin, it is lower in blood content, and depends instead on the lymph system for oxygen. The dark muscle is generally higher in fat and is more easily oxidized.

INTERNAL ORGANS

The internal organs of fish are somewhat different from those of higher animals. The digestive tract is designed for periods of feast or famine. There is no mastication of the food. The stomach is either very stretched or essentially nonexistent. The intestines are relatively short. The digestive enzymes are synthesized in the stomach lining, intestines, and pyloric caecum. Unique to fish, the pyloric caeca is a source of various enzymes and has been used by scientists to prepare the enzyme that is responsible for converting trimethylamine oxide to dimethylamine and formaldehyde. It is not clear what the relationship of this enzyme in the pyloric caecum is to the same enzyme function found in the muscle tissue.

The kidneys of most fish are not distinct organs; rather, they are an area of blood accumulation beneath the backbone. This area must be removed carefully if a longer shelf-life is to be obtained, usually by "brushing." To get to the kidney, the swim bladder (see next paragraph) must be removed first.

Many fish are able to regulate their buoyancy using their swim bladders. This is a specialized organ made of collagen and a guanido compound similar to the material found in fish scales. Both the swim bladder and the scales can be used for pearl essence, that is, the "sparkle" in nail polish. The swim bladder removes gases from the bloodstream and adjusts the size of its "balloon" to account for the pressure/density changes in the water column as the fish moves through the water column. When a fish is removed from deep water, the swim bladder may expand rapidly. It may burst or it may cause the other organs of the fish to be forced out through the mouth. This can be helpful in cases where whole ungutted fish might be marketed, since it generally leads to an emptying of the gut tract.

The cartilaginous fishes such as shark have only one ovary. They produce a few large eggs, which have a horny shell. When these were found by marines they were referred to as "mermaid's purses."

POSTMORTEM CHEMISTRY OF FISH

Following death, a number of changes occur in all animals. The generally complex and tightly controlled biochemical system breaks down. As ATP is depleted, the ability of the actomyosin linkages to separate is decreased. Finally, the muscle goes into *rigor mortis*, the stiffening of death, which is caused by the formation of permanent linkages between the actin of the thin filaments and the myosin of the thick filaments.

From the moment of death, the process of *rigor mortis* takes place in about 2 hours in fish. It usually starts in the tail because the tail has the smallest heat capacity. To test for *rigor mortis* in small fish, hold the fish upright with a straight-up clip at the tail; see how much bending occurs due to gravity—"onset" means that the fish no longer bends below the clip; in pre-rigor it bends below the clip. Another way to measure rigor may be the R value, which is the ratio of adenosine to inosine compounds; it is measured as the ultraviolet spectral absorbance at 258 nm divided by the absorbance at 250 nm. If the fish is still warm at that point, that is, it is in a warm environment, there might be a contraction during the onset of *rigor mortis* that would tear the flesh. If the fish has been packed in ice and has assumed an odd shape (i.e., it looks exactly as it was packed) it should be left that way, that is, not cut, until rigor resolves itself and the fish muscle can then be straightened out without tearing. Warm water fish undergo the "cold-shock reaction" shortly after being placed on ice. The fish stiffens, but—unlike "cold shortening" in mammalian muscle—the fish muscle is not really contracting. Apparently the metabolism is increased at these lower temperatures due to ATP degradation, lactic acid build-up and a drop in pH. Freeze–thaw cycling can reduce the amount of at least one cofactor required for the dimethylamine and formaldehyde reaction.

These chemical changes can take place even if the fish muscle is slightly frozen (-2 to $-4°C$; 28 to 25°F). If a fish is frozen after its death but before *rigor mortis* sets in, the process of rigor development can occur without inducing thaw rigor. Thaw rigor is the massive contraction occurring prior to *rigor mortis* as a result of a pre-rigor fish's being thawed quickly. The ATP and calcium levels are very high, and the muscle is very tough and dry.

The gaping phenomenon described earlier also depends on the rigor state. Post-rigor fish are most likely to gape; fish in the midst of *rigor mortis* sometimes gape; pre-rigor fish rarely gape, probably because they have yet to reach their ultimate pH. In all cases, a higher rigor temperature seems to encourage gaping. Whole fish gape more often than fillets do, possibly because they offer more resistance to change in the sarcomere length; the fibers put more strain on the myocommata.

The resolution of rigor—the return of pliability to meat following rigor—is believed to involve various proteolytic reactions that occur at points in the sarcomere other than the A band, that is, the region of thin and thick filament overlap. The calcium-activated protease that seems to act at the junction of the I band and the Z line in higher animals has been postulated to have an important role in rigor resolution. For further information about muscle and meat biochemistry, the book by Pearson and Young (1989) of the same title is recommended.

The lowering of the postmortem pH leads to a reduction in the water-retention capabilities of the fish muscle. Therefore, the fish with the lowest ultimate pH also has the greatest chance of dripping, that is, losing moisture. The loss of moisture may lead to the organoleptic condition known as chalkiness. The loss of water leaves the remaining flesh very dry. These water retention properties of fish flesh are often referred to as the fish's "water holding capacity" and can be measured in one of two ways. The first involves expressing moisture from the material as it normally exists by the use of an applied force such as centrifugation or a hydraulic press; the resulting liquid is called expressible moisture. Another technique involves adding a great deal of excess aqueous solution to the flesh and then measuring the water held by the insoluble phase; this is called the water binding potential. These two different methods for measuring water holding capacity respond differently to pH and the presence of various ions (Regenstein 1984).

Postmortem changes of fish lipids greatly affect the final product. Oxidation of lipids leads to rancidity. The odor of rancid fish fats has been referred to with terms such as oily, fishy, seaweedy, and paintlike. The cause is usually aldehyde- or ketonelike compounds formed from the shorter-chain fatty acids. Hept-cis-4-enal is often measured as a typical member of this group. The additional fat in the high-fat fish is made up of triglycerides; their membrane fat needs are no greater than those of the low fat fish.

Phospholipids such as lecithin are generally higher in polyunsaturated fatty acids and are an integral part of the cell membranes. They seem to play an important role in creating the "warmed-over" flavor of red meats, that is, the flavor associated with meats that have been precooked and reheated. This problem must be considered in the development of "already-cooked" fish products, particularly fried products.

Free fatty acids have been associated with protein cross-linking and other negative changes in fish such as texture breakdown and more rapid lipid oxidation.

It appears that bacteria can penetrate into fish flesh along the collagen fibers if the temperature is above 8°C (46°F). Below this temperature bacterial spoilage remains a surface phenomenon. Some proteolytic enzymes of bacteria can break down collagen.

REFERENCES

Aitken, A., I. M. Mackie, J. H. Merritt, and M. L. Windsor. 1982. *Fish Handling and Processing*. Edinburgh, Scotland: Her Majesty's Stationery Office.

Love, R. M. 1970. *The Chemical Biology of Fishes*, Volume 1. New York: Academic Press.

Love, R. M. 1980. *The Chemical Biology of Fishes*, Volume 2. New York: Academic Press.

Love, R. M. 1988. *The Food Fishes: Their Intrinsic Variation and Practical Implications*. London: Farrand Press.

Pearson, A. M., and R. B. Young. 1989. *Muscle and Meat Biochemistry*. San Diego, CA: Academic Press.

Regenstein, J. M. 1984. Protein–water interactions in muscle foods. *Proc. Recip. Meat Conf.* **37**: 44–51.

Regenstein, Joe M. and Carrie E. Regenstein. 1984. *Food Protein Chemistry*. Orlando, FL: Academic Press.

13

Aquaculture is the controlled growing of specific aquatic life forms; it can be thought of as the aquatic equivalent of agriculture. The term "mariculture" is sometimes used when saltwater fish and shellfish are involved. The concept of aquaculture actually includes a broad range of possible activities with a concomitant range of control over the process. In the most controlled cases, the goal is to raise a species of fish from conception to death in every generation indoors in recycled water. A far less controlled case might include obtaining from the wild the starting stock or just the eggs and sperm. In ocean farming and shell stocking, the hatchery-reared animal is returned to the natural, or wild, environment. Different species demonstrate varying needs in reference to successful survival in controlled aquacultural situations. In an unusual example it has been found that fish cultured at high altitudes may be harmed by ultraviolet radiation, especially young fish; the damage to the skin can result in both fungal and bacterial infections. In some cases it is necessary to provide shade for the fish. The turbidity of the water will also have an impact on fish growth. Some fish actually prefer slightly lower water quality.

An important distinction between land-based agriculture and water-based aquaculture is the emphasis on environmental quality. Monitoring water quality parameters and understanding the related water chemistry is extremely important. The oxygen content must be high enough to support fish growth, and the ammonia content, the metabolic waste product of fish, must be low enough. Factors like pH, temperature, and alkalinity all affect these factors and must be properly accounted for in operating an aquaculture facility. As temperature increases, gas solubility decreases but metabolic rates increase. The species selected must match the available environmental conditions. However, the species selected must also have the required marketablity so that the project can be economically successful.

Diets of Aquacultured Fish. An aspect of the controlled aquacultural environment that has received particular attention recently is the effect of controlling the fish's diet. The very composition of the aquacultured fish can be changed by this method. For example, many fish need omega-3 fatty acids in their own diet; increasing this dietary component in an edible fish can ultimately enhance the human diet.

Feed intake of fish is determined largely by metabolic rate, which is directly proportional to water temperature. Reduced glutathione may also stimulate the feeding response. In any case, odor does not seem to be significant to the fish: rejection of repugnant diets is demonstrated by the fish's spitting the food out of its mouth. Certain water-soluble flavors such as some amino acids and the compound betaine, however, seem to aid in making feeds acceptable.

In making extruded fish feeds, it is important that the starch component be cooked beyond the granule rupture stage so that the starch can be digested by the fish. The extruded product should be in discrete particles which hold together: the bulk density must be appropriate for the desired use, that is, the particles must usually have a more porous internal structure if they are expected to float rather than sink; the particles should absorb water without falling apart and continue to either sink or float as needed. Cooking is more uniform with smaller grinds. Fat tends to increase the density of each particle; it also helps to maintain the porous structure when added after extrusion.

Floating feeds force the fish to come to the surface so that they may be observed by the aquaculturist. However, the currents and wind may cause the most rapid drifting away of the feed. Sinking feeds are often preferred by the fish and may be easier to manufacture.

Some work is currently being done to develop diets that minimize the waste generated in the water. Denmark has legislated that the feed efficiency must equal 1 or better, that is, one pound of feed for one pound of weight gain, to minimize pollution of the area directly under and around fish pens. There are reports that the waste that accumulates underneath a fish pen leads to a change in the ecology and possibly to undesirable anaerobic conditions.

Benefits of Aquaculture.   It is currently estimated that 10% of the total world fish catch benefits from some form of aquacultural handling. The lower figure of 3% for the United States' total catch probably reflects both an abundant coastal stock and an absence of strong pressure, or leadership, for more efforts to develop aquaculture. At this writing, Idaho trout and Southern catfish are the only successful large-scale aquacultural finfish in this country. Other species are being developed around the country, including striped bass, tilapia, sturgeon, red drum, shrimp, clams, oysters, and abalone. Alaska, because of the importance of wild salmon, has outlawed fish aquaculture.

There are approximately a thousand catfish farms in the United States, most of which are in Mississippi, Alabama, and Arkansas. The ponds generally vary in size from 8 to 260 hectares (20 to 640 acres). In recent years, as it appears that water availability will become more limiting, the trend has been to use smaller ponds with a more intensive stocking level. In 1985 the total United States catfish production was about 86,900 tons. The farmer receives about 75 cents/lb at the farm for the whole fish, which translates into $1.40 for whole iced fish, including 15 cents for the freezing. Although a fine product can be produced, further marketing opportunities must be developed.

Recent overfishing of natural stocks, both high-priced species and the traditional cod, haddock, and flounder species, may serve to further finfish aquaculture efforts in the United States. Salmon aquaculture is one such opportunity. However, the economics of growing salmon in the United States versus importing it from other countries has not yet been determined. In Oregon, smolts (small fish ready to release into the marine environment) have cost about $0.30 to $0.40 for 35–40-gram coho, and $1.50 to $2.00 for Atlantic salmon smolt. These high prices for starting materials can be a deterrent to aquacultural development. Such costs would also limit finfish aquaculture to the pattern of shellfish, such as clams and oysters, whose premium price has encouraged these types of aquaculture in the United States for many years.

Goals of Aquaculture.   The goals of aquaculture around the world are not unlike the goals of the "wild" fishing industry:

- *Natural stock improvement:* Returning these animals to the wild at some stage to replace the natural stock.
- *Sport:* Enhancing the sports fishery by placing fish in various waterways for sports fishermen to catch.

For these first two goals, aquaculture has played an extremely important role. The sports fish industry has supported state and federal fish hatcheries, which, in turn, have reared commercially important species such as trout and salmon. The federal salmonid program is run by the United States Department of the Interior's Fish and Wildlife Service. In addtion to serving the sports fish industry, these activities have provided much useful information about successful aquacultural procedures.

- *Bait:* Providing small aquatic animals that are used to catch other fish. This is an under-rated service whether it is provided aquaculturally or in the wild. For example, the catching of worms for bait is an important economic component of the Maine fisheries. Arkansas is the major bait-producing state in the United States. Minnows, crayfish, and suckers are important in the more northern states.
- *Hobby uses:* Providing the many varieties of fish, often very colorful, that

are sold to people with household aquariums. This is a surprisingly lucrative trade, especially with fish native to the warmer water countries.

- *Pollution testing:* Because fish are very sensitive to various pollutants, they can be used as excellent signals of water quality problems. This is especially true of fish maintained in captivity. Heat pollution, that is, an excess of hot water dumped into the natural water supply, has become a concern because of its potential impact on aquatic life.

- *Recycling waste:* Utilizing already-occurring resources such as the heat from power plants: many fish farms are located near power plants to take advantage of their waste heat. An interesting related example in Hawaii is the use of ocean thermal energy gradients, the approximately 22°C (72°F) natural temperature gradient between the surface and the bottom of the ocean, to generate the electricity needed to power an aquaculture venture. The water from the ocean bottom also seems to be higher in trace nutrients such as nitrate, nitrogen, and inorganic phosphorous, thus permitting a better algal or single-cell protein growth, which can then be used as feed for fish.

- *Industrial commodities:* Raising certain animals for specific compounds that they may synthesize. For example, it has recently been shown that some fish seem to have a very specific chemical defense mechanism against sharks. The isolation of the chemical(s) involved may be worth pursuing.

MARICULTURE

The mariculture of various shellfish, such as mussels, oysters, and clams from spat, that is, the young larval form, is carried out on the ocean bottom, on racks, or on other forms of suspension such as ropes, in the water column. Many of these molluscan animals are called "filter feeders" and are able to obtain their feed from the particulates floating in the water column. As long as the water keeps moving and enough organic feed is available, these animals do not need to be fed, although feeding will generally speed growth. For example, abalone are cultured with algae indoors: at 6 months they are removed from the algae tanks and placed outside into tanks for 18 months to reach market size, that is, 8.25 cm (3.25 inches) in diameter, yielding a one-ounce steak.*

For those animals that can attach themselves to other materials such as

---

* Many aquacultured products tend to be marketed on the small side. The longer animals are kept, the greater the risk and cost of a loss. However, in the long term, aquacultured products are going to have to be size-optimized by the market, not by the growers' needs.

shells or ropes, suspended raft cultures become possible. The meat yield for mussels has been so successful using rope cultures that some countries' aquacultured supply has outrun the market; for example, excess product is used for animal feed in Mexico. A small quantity of mussel meal as a chicken feed can heighten egg yolk color. However, the natural orange/reddish pigments associated with mussels can yield eggs with an unusual orange-pigmented yolk if mussel meal is a significant component of the feed. A redder egg is considered desirable in some countries, particularly with duck eggs.

Shrimp Aquaculture.   The most popular shellfish in the United States is the shrimp. The demand has been estimated at about 300,000 tons per year. This has encouraged efforts to build hatcheries for post-larval shrimps, which used to be caught wild. As long as the industry is dependent on obtaining larvae from the wild, the potential production each year will be limited by the natural forces limiting this yield. Because the stocking densities in the ponds are low, the yields of adult shrimps are only 605 kg/hectare per year in Latin America and 400 kg/hectare per year in Asia. In recent years more intensive shrimp aquaculture systems have been developed that give higher yields per hectare. Hawaii would have a particular advantage in shrimp aquaculture in the United States because of its warm climate: farms could get almost three crops of shrimp per year. If the continental United States is to develop shrimp aquaculture, it will probably be with an intensive system that optimizes feed, energy, and labor costs.

The baby Mexican white shrimp, *Penaeus stylirostris*, needs a 23-week grow-out period. Grow-out is at ca. 57°C (75–76°F). It then takes 105 days to grow to 31–40 count per pound tail weight. The feed conversion ratio is three to one. The shrimp are not fed for the 2 days prior to killing so as to clean out their gastrointestinal tracts. Color is controlled by the feed. Shrimp aquaculture has been most successful in Ecuador and China.

Penaeus orientalis is a white shrimp that is grown in ponds and imported from China. Ecuador raises the white shrimp *Penaeus vannamei* and the slightly less valuable blue shrimp *Penaeus stylirostris* in shallow ponds. Ponds are a maximum of 2 m (6 ft) deep and use brackish water, 15–25 parts per thousand salinity. Post-larval shrimp, captured from wild stock, are first grown in a nursery pond and then introduced into the grow-out ponds. Water temperatures of 30°C (86°F) allow harvest after about 120 days. Pond densities range from 20,000 to over 40,000 post-larval shrimp per acre. Three crops may be harvested per year. Fertilizers and feed may be added in the more intensive operations.

Ecuadoran shrimp aquaculture has suffered recent reverses: the supply of wild larvae was greatly reduced, possibly because of cold water temperatures, a shortage of rain, or overfishing. A virus has affected the white shrimp and no solution to the problem currently exists. Furthermore, Ecuador's extensive system of shrimp farming yields only 360 kg/hectare even with three crops per year. Although the seed stock was trapped initially in nearby mangrove areas, this supply is not sufficient and hatchery capabilities must now be developed.

Lobster Aquaculture.   Lobster have been of particular interest in recent years. Lobster have a high market value but present special problems because they are marketed live.

Betaine, a dipeptide, is found in most crustaceans in the range of 0.1 to 1.4%. The free amino acid content of crustaceans is also high; glycine often constitutes more than 1% of the species. These compounds are water-soluble and may serve as attractants for feed and artificial bait.

Canner lobsters, that is, those legally caught in Canada that would not be legal in the United States, can be grown out to United States market size using a series of stackable isolation plastic floats, each with a few separate sections so each lobster remains isolated. The filter system for the recirculated water involves ozone sterilization, mechanical filters, and other filters. It takes about 6 months to grow out the lobsters using a feed consisting of crab and fish scrap, wheat, and a vitamin premix. The feed must be pelleted as the lobsters do not always eat immediately. The cone-shaped bottom of the cylinders permits the removal of sludge.

There are a few specific techniques that enhance lobster aquaculture. One trick is to use an electrical current to get the males to release their sperm sacs. These sacs are normally transferred to the female during the mating process. Another method to assist reproduction is to create a light cycle of 80 short days followed by 120 long days, causing the female to "egg out." Controlling the egging-out season permits aquaculturists to place the egg-laying process between molting cycles. This is preferable to the "natural" timing in which molting generally interferes with egging out. Finally, lobster aquaculturists manipulate water temperature to shorten the usual year from the time of egg laying to hatching. This technique also yields larvae at any time of the year.

POLYCULTURE

The underlying concept of polyculture is that different fish occupy different ecological niches in a pond. If the right assortment of fish and plant life can be put together, then the total productivity of the pond is maximized. This goal is in contrast to the goal of "monoculture" where aquaculturists try to maximize the yield of a single species, and therefore, usually have to leave all of the other available ecological niches uninhabited. The yield of the highest-value species obtained may be decreased in a polyculture system, but the total return from the pond may well be greater if the other niches are well utilized. Polyculture is especially valuable in countries for which water is the most limiting resource, such as in the deserts of Africa and the Middle East.

Much of the research on polyculture has been done in Israel. In one example, three different types of fish are grown in one pond: carp (benthophagic, bottom feeders), tilapia (detriophagic, which helps control weeds) and silver carp (phytoplantophagic, they eat the small plankton in the water column).

The fish raised in these ponds are all considered excellent eating fish. Carp is a hardy fish with a long tradition of both aquaculture and wild catching. It is used in Central European and Jewish cooking, particularly as an ingredient in gefilte fish, originally a carp stuffed with a fish stuffing. The fish stuffing is now cooked on its own as essentially well-ground fish, egg, matzo meal (an already baked cracker crumb), and seasonings boiled for about an hour. Carp is also popular in the Chinese ethnic market. To date, however, local Israeli markets have not been able to sell all of the aquacultured carp. On the other hand, gefilte fish always draws a premium in the United States when sold in Jewish neighborhoods. Regardless of the anticipated market, carp requires special attention at least for the last few days before slaughter. If the waters contain any blue-green algae, the fish can pick up a "muddy" flavour.

The New Testament refers to tilapia as the fish that St. Peter caught in the Sea of Galilee. Christ used it to feed the multitudes. There is also evidence from the pyramids that tilapia was indeed grown in that era. The likely species was *Oreochromis mossambicus*, but current efforts at tilapia aquaculture have been moved forward by the use of the Nile tilapia, *Oreochromis niloticus*, or, in areas where cold-tolerance is important, the blue tilapia, *Oreochromis aureus*.

Tilapia in its many forms is one of the three most common finfish species used for aquaculture worldwide. The others are milkfish, which is very popular in the Pacific basin, especially the Philippines, and carp, which has been grown in the Orient for many years.

Tilapia can be grown so that they gain up to 2 grams of body weight per fish per day. They are fairly tolerant of poor water conditions, but require approximately 10°C (50°F) water in which to grow. The 10- to 20-ounce fish (280–560 g) can be filleted to yield two boneless fillets of 2–5 ounces (56–140 g) each; the meat is very pleasant, firm, and mild.

Since the males grow faster than the females, it pays to raise them separately or exclusively. Tilapia have also been aquacultured in Arizona and Idaho as well as in the Philippines, where pig manure has been used as a fertilizer, apparently with no resulting human health problems.

A variation of this polycultured pond incorporates the growing of shrimp (*Macrobrachium*). However, the more complex the pond arrangement, the greater is the need to carefully monitor the production of food by the plant components as well as the relative masses of each plant or animal component. This requires a more technically sophisticated operation. The market for the shrimp grown in Israel, where it is not sold because it is not kosher, is in Europe. Like Ecuador, which raises the same species, both countries must rely on a wild seed stock that has become severely limited.

SALMON

Much work has been done in recent years to commercialize aquacultured salmon, especially Atlantic salmon. The fjords of Norway are ideal since "penned" salmon thrive in the protected bays where the movement of the water is sufficient to permit the waste produce by the fish to be moved out naturally. It is a challenge to farm salmon: the net system must not foul easily; it must offer protection from seals and waves, and—to some extent—winter temperatures. Most salmon need temperatures just above freezing, to about 12°C (54°F) for optimum growth and spawning. The fish should be moved via piping rather than by handling since the loss of even a few scales can be lethal. Two diseases must also be avoided: bacterial toxin in the blood, which can be treated; and bacterial kidney disease, for which there is no known treatment and which can be transmitted through the eggs.

Norway is also very motivated economically because it is a small country with a lot of coastline. Although the Norwegian fjords protect the salmon pens from the worst ravages of weather, the mixtures of cold temperature and wind in any northern climate can lead to the superchilling of the water, which ultimately kills the fish. This creates a paradox: salmon would thrive if grown in a submerged culture, but they must obtain air to fill their swim bladders, by "gulping." They can be trained to respond to a light for feeding.

Salmon has also been aquacultured on the west coast of Scotland, an otherwise underdeveloped area. Scottish marketing efforts can take advantage of the country's excellent reputation for wild salmon products, both fresh and frozen.

Efforts at salmon pen farming have occurred in far-northern Maine and in British Columbia and New Brunswick, Canada. Other countries involved in pen-raising salmon include Chile, Iceland, and Ireland among others. The list grows.

Some of the Norwegian, Scottish, and Chilean salmon is flown into the United States as fresh product, particularly in the winter months when there is no fresh wild salmon available and it is easier to arrange for air shipping. Chile has the additional advantage that it is summer in the southern hemisphere while it is winter in the northern. Other important markets are in Europe and Japan.

The efforts in British Columbia, Maine, and New Brunswick might eventually compete by saving on shipping costs if they can overcome the higher labor costs and capital requirements present in North America. A number of the farms in British Columbia are in financial difficulty, and Ocean Products, the largest farm in Maine, was recently sold after it ran into financial difficulty.

North American salmon aquaculturists benefit from another advantage: the United States FDA has recently determined that some of the imported fish contain levels of tributyltin (a toxic compound used to prevent the growth of barnacles and other organisms on boats, nets, etc.) of 0.28–0.90 µg/liter. The compound is not heat-denatured and is considered one of the most toxic compounds known. It is toxic at levels of 5 parts per trillion. It is used to treat netting in saltwater pens like those found abroad. Nevertheless, current farmed salmon predictions are 90,000 tons from Norway (about 550 farms) and 20,000 tons from Scotland. New Zealand, Chile, Iceland, and Ireland are also trying to enter this market. The annual United States catch of wild salmon is 6–7 million tons; salmon ranching, that is, release and recapture, may add more although the success of this technique is questionable.

The traditional diet for aquacultured salmon has been high-protein animal-based feed such as herring with fish meal. However, recent data from the Tunison Laboratory of the United States Department of the Interior (Cortland, New York, and Hagerman, Idaho) suggest that properly formulated plant-based feeds work as well in diets for salmonids (Smith et al. 1988). Taste panel tests done at Cornell indicated that the flavor of frozen trout fillets was not changed by a switch in diet. However, a move to plant-based diets may also lead to a lower omega-3 content in the final fish. This might be a negative from the consumers' point of view, even if it would make the fish more stable during storage.

The commercial feed-to-weight ratio for salmon is about two pounds of feed per pound of gain; the dry feed costs about $0.25 per pound, moist pellets cost a little more. Current research suggests that the diet of Atlantic salmon influences its final fatty acid composition, for example, menhaden oil yields a different final lipid profile than does soybean oil.

TROUT

Trout are grown in Idaho because of a favorable natural resource: large quantities of very clean water come out of springs high above the Snake River at almost the ideal temperature of 15°C (59–60°F). This water can usually be run through the raceways without any pumping and, after minimal treatment, can be released back into the Snake River. However, this water is running out and government standards for the quality of the released water are increasing, suggesting that water-management practices in the Snake River trout-growing area will be changing in the coming years.

Other locations are not similarly blessed. However, trout have been raised in brackish water, and even in sea cages. In an experiment in Great Britain, trout pounds were oxygenated to increase the stocking density; the British Oxygen Company reported technological success, but lack of economic viability. In another technically positive but economically negative project, the Kroger Company attempted to use high silos for raising trout; the facility was ultimately donated to the Arkansas Fish and Game Commission.

Sometimes smaller niche markets can be developed. For example, a small commercial trout industry in New York's Delaware County has been able to use aquaculture to serve the local population. This has been aided by an additional marketing effort in the adjacent resort area of Sullivan County (the Catskills). The growth in trout rearing in Western North Carolina also represents such a niche opportunity.

Many aquaculturists have undertaken breeding experiments in order to select appropriate stock of particularly large fish. Of course, consumer preferences are critical for commercial success. For example, British and apparently Japanese consumers prefer female trout; efforts are therefore underway to influence the trout's sex programming at birth to satisfy this special demand. If trout sperm is treated with ultraviolet light, only females are produced. Work is also being done with a triploid trout.

Nutritionists have noted that longer-chain fatty acids seem to help in the absorption of the color additive canthaxanthin in trout, and that the addition of extra vitamin E to the trout seems to increase both canthaxanthin and polyunsaturated fatty acid levels in edible trout.

Note: Currently the legality of the use of pigments in the feed of fish consumed in the United States is questionable. Direct addition of the pigment into the feed is certainly not legal, although the Boston office of the FDA seems to be the only part of FDA that is enforcing these laws. There is also a question of whether such usage must be labeled at retail. The use of pigment derived from yeasts or other microorganisms intentionally grown for pigmentation also seems to be questionable. The feeding of pigmented shellfish wastes such as crab and shrimp shell is legal.

GADOIDS

Norway has begun a project to grow cod by aquaculture, using enclosed saltwater ponds within the fjords. In Canada there is an effort to take early summer trap-caught cod and transfer them to grow-out net pens for harvest in the late fall when prices for cod are higher.

FLATFISH

Efforts are taking place in many European countries to raise various flatfish. For example, turbot larvae need 20°C (68°F) while the juveniles weaned at 3 months take about 36 months at 14–18°C (57–64°F) to reach 2 kg (4½ lb) market weight. Turbot is being grown in Spain and halibut seems close to being commercially possible.

SEAWEED

The growing of seaweed is a specialized subarea of aquaculture found in the Orient, where the indigenous population attributes many medicinal properties to this product. Western nutritionists do agree that seaweed has many valuable trace minerals. Seaweed aquaculture has been so successful in China that it is cost-effective to fertilize the ocean, actually, large bays, with chemical fertilizers that are bought from Western suppliers.

The international seaweed trade includes dried brown seaweed (some salted), wet brown seaweeds, and the red seaweed *Porphyra*. McHugh and Lanier (1984) lists seaweeds for human consumption:

Red seaweed: *Porphyra*
Brown seaweed: *Hizikia, Laminaria, Undaria*
Green seaweed: *Caulerpa, Ulva, Enteromorpha*

Most of the dried brown seaweed is *Undaria pinnatifida*, known as "wakame" in Japan. Kombu are brown algae, mostly *Laminaria*, and are often referred to as "kelp." These products can be boiled as a vegetable, used in soup stock, used to season rice dishes, or eaten dried as a snack. When wakame is used in miso-soup, a popular dish in Japan, the salted, wet form replaces some of the dried form. Another form of seaweed is *Undaria*, supplied to Japan as "salted mustard." Among the red algae, Irish moss is used for carageenan extraction and is probably *Chondrus ocellatus* and/or *Gigartina tenella*. *Gelidium amansii* is used for agar. The *Porphyra* (purple laver) has a high protein content (25–35%), and is used to wrap sushi. Most of the green algae produced is "green laver," that is, *Enteromorpha* species (McHugh and Lanier, 1984).

Seaweed is a source of key food ingredients such as alginates and carageenans. The alginates come from brown algae such as *Macrocystis pyrifera*, one of the fastest-growing plants known. In the United States, extensive harvesting is done off the coast of Southern California by the Kelco company. A "grass-cutter" is used about 4 feet under the water surface; then the kelp is dried and processed. Like grass, seaweed grows back quickly and can be reharvested many times.

Kelp.   Kelp is also abundant in Mexico. Hernan Mateus (1975) became interested in processing *Macrocystis pyrifera* for its alginate content as a cottage industry in Baja, California; he also wanted to use the remaining residue as a poultry feed. There were problems to overcome. Potable water is very limited there and energy costs are high; the kelp's high mannitol content has to be removed because it would cause diarrhea in chickens.

Mateus showed that the alginic acid could be extracted without the traditional cooking step using undried kelp. He used seawater rather than freshwater in the extracting solutions to obtain a yield of alginic acid approximately 85% of the maximum textbook value for the process that involved drying, heating, and so on. Mateus' willingness to depart from tradition and use a new technology helped him to lower production costs and greatly improve the feasibility of the entire project (Mateus, 1975).

RANCHING

In aquaculture ranching, the fry are raised to fingerling size and then released into the wild. By this point, the fish are supposed to have been imprinted with the sense of their home base. It is hoped that after 3 or 4 years at sea, for example, salmon ranched in the United States northwest will return home—as would wild salmon—and be captured by those who first released them.

Although ranching operations are predicated on a low (2%) return of the originally-released stock, even this return is not always achieved. In a recent case, the aquacultured salmon found their way back to the main stream from which they started out, but they did not turn off into the small tributary stream to return to the operators who had released them.

Current research indicates that radioimmunoassay methods may be used to measure thyroid hormone levels which indicate smoltification and readiness to migrate, that is, the ability to function in saltwater. Moreover, "schooled" salmon are currently being exposed to predatory fish through a window to teach them to avoid predators; this has improved their survival rate.

AQUACULTURE VERSUS AGRICULTURE

Although aquaculture development in the United States is many years behind that of agriculture, the potentials for obtaining and managing aquatic resources are significant. Indeed, many observers believe that raising fish by aquaculture has advantages over the raising of terrestial animals for the following reasons.

- *More efficient feed conversion rate.* The most efficient land animal is the chicken, with a feed conversion rate of about 1.8; that is, it takes 1.8 pounds of feed to grow one pound of chicken. Note that no adjustment is made for differences in the moisture content of either the feed or the animal. This figures rises to eight or nine pounds of feed per pound of animal for feedlot beef. Although the protein content of fish diets is much higher than those of other animals, and, with it, the cost of the feed, the feed conversion rate for fish can be as low as 1.0; that is, every pound of feed yields a pound of live fish. There are three primary reasons for this difference.

  1. Fish are buoyant and water offers limited friction, so fish use less energy to move around and to maintain themselves at rest. They do not have to carry around a heavy skeleton!
  2. Fish are cold-blooded and therefore do not need to use extra energy for heat production.
  3. Fish recover the full 5.4 calories per gram of energy found in proteins when they metabolize protein feed. Their system breaks the protein down to ammonia that can be excreted. The ammonia is toxic to fish and must not be allowed to accumulate in the water. However, most land animals, including humans, recover only four calories during protein metabolism because they expend 1.4 calories to remove ammonia by converting it to urea before it can be excreted.

- *Reproductive efficiency.* One fish can lay from 10,000 to many millions of eggs each year. Thus, maintaining one female for one year, that is, feeding her and caring for her, can yield many millions of offpsring. Agricultural animals are far less efficient. A commercial production hen can lay a maximum of 300 eggs per year and the birds used as breeders, that is, reproductive stock, lay many fewer eggs per year. Turkeys lay even fewer eggs, a limitation which has become important in current efforts to expand turkey production. One cow can have only one or two calves each year. Relatively fewer male fish are also required than male agricultural counterparts.

Of course, there are still important processes in which agriculture leads aquaculture; aquaculture can learn from these experiences, for example, in diet formulation and breeding programs. Major areas where agriculture scores over aquaculture are the following:

- *Control of wild stock.* Greater control is possible of aquacultured fish than wild stock throughout the life cycle. Aquaculture creates the ability to harvest fish as needed and control the slaughter conditions. However, terrestrial animal slaughter has been far better controlled than the aquatic counterpart. Aquacultural development must include improved slaughter and handling procedures.

- *Domestic marketing.* The marketing of aquacultured products has at times not kept pace with supply in this country. Both trout and catfish have found themselves in overproduction situations because of a failure to fully exploit the United States market. Salmon may face this problem in the next couple of years.

- *By-product utilization.* Researchers are trying to seek new and profitable uses for fish by-products, for example, catfish mince. Such efforts have been proven successful in the enhancement of the meat and poultry industries in the United States. Mince was discussed in detail in Chapter 9.

PERSPECTIVES

For aquaculture in the United States, the operative word is "potential." Its success will depend on more than the availability of appropriate water sources. It will depend on many legal, political, social, and economic conditions. Although American citizens have increasingly demanded improved alternative uses of land and water, change is historically slow, especially when governmental leadership is necessary. Various states have recently developed state-wide aquaculture plans, that is, guides describing what needs to be done to make aquaculture viable. In some ways, these reports are discouraging: numerous recommendations must be fulfilled by the government before the investment support of the private sector will be attracted. And, of course, the reports do not even list the innumerable scientific and technical concerns that must be studied and resolved in a timely fashion. Even more discouraging is the fact that although most of these reports were issued a few years ago, very little has changed!

Regardless of aquaculture's potential, the wild catch of many fish species will not only continue but will thrive. The wild catch will generally remain more economic than aquaculture in the United States until the cost of vessel fuel or the effect of overfishing or some unknown factor disrupts the current economics of utilizing aquatic resources. The only areas in which aquaculture might currently compete with the wild catch will be for commercially important species such as cod, halibut, haddock, flounder, and ocean perch, which have been overfished.

This is not necessarily the case in less-developed countries, where proper conditions for developing aquaculture as a major food source do exist: (1) The limited availability of alternative sources or of a fresh or frozen fish distribution system encourages aquacultural development. (2) "Farmed" freshwater fish can be grown near the site of consumption, sometimes impossible for a wild-stock fishery. (3) Cheap land may be available that is inappropriate for other agricultural purposes. (4) Low costs for labor can be a great help. Conditions like these have encouraged aquaculture in countries such as the Philippines and China, where fish productivity has actually surpassed that of more developed countries.

A final caveat: aquaculture is not just the growing of fish; its real goal is to provide consumers with fish to eat or to meet other appropriate consumer needs. The post-harvest handling, distribution, and marketing of aquacultured fish are critical to the successful use of the harvest. The aquaculturist's investment is great; it is also vulnerable. If the price collapses for a wild-caught species, the fisherman can simply stop fishing. But, like his agricultural colleagues, the aquaculturist has the animals on hand and must continue to feed and care for them.

The aquacultural process must be perceived as a complete economic undertaking from initial investment to ultimate profit. Only in this way might there ultimately be a profit for producers and consumers alike. Table 13.1 gives some current FAO aquaculture figures.

Table 13.1 Leading Aquaculture Countries.

Country	Value ($US 000s)	1988 Production (tonnes)	Key species or species groups
China	7,950,523	6,658,686	Carps, molluscs, shrimps
Japan	4,571,887	1,424,832	Amberjack, molluscs, algae
Taiwan	1,203,088	300,975	Eel, molluscs, shrimps
Philippines	720,294	599,464	Milkfish, shrimps, algae
USA	608,858	402,757	Catfish, Pacific salmon, molluscs
USSR	605,580	364,783	Carps
Norway	589,268	89,410	Atlantic salmon, rainbow trout
Ecuador	580,625	75,631	Whiteleg shrimp
Korea RO	550,972	900,292	Molluscs, algae
Indonesia	550,781	481,370	Carps, milkfish, shrimps, algae
Korea DPR	485,500	831,000	Molluscs, algae
France	471,022	222,339	Rainbow trout, molluscs
Vietnam	355,680	146,700	Freshwater fish, crustaceans
India	352,578	437,130	Freshwater fish, crustaceans
Spain	288,515	268,758	Rainbow trout, molluscs
Thailand	261,895	192,582	Crustaceans, molluscs
Bangladesh	209,987	154,834	Freshwater fish
Italy	191,817	97,000	Rainbow trout, molluscs
UK (Scotland)	119,601	22,420	Atlantic salmon
Romania	95,000	38,000	Freshwater fish

Courtesy of FAO.

REFERENCES

Mateus, H. 1975. Studies on the marine brown alga *Macrocystis pyrifera*. M.S. Thesis. NY: Cornell University, Ithaca.

McHugh, D. G., and B. V. Lanier. 1984. Korea's edible seaweed trade. *Infofish Marketing Digest* No. 3, 17–19.

Smith, R. R., H. L. Kincaid, J. M. Regenstein, and G. L. Rumsey. 1988. Evaluation of select strains of rainbow trout (*Salmo gairdneri*) fed diets composed primarily of plant or animal protein. *Aquaculture* **70**: 309–321.

14

Human Nutrition and Public Health Concerns

Is fish good for you? The popular press in the United States would certainly have readers believe that it is. Countries with fish diets boast the highest average life span: approximately 74.2 years for men in Japan and approximately 79.7 years for women in Iceland.

Let us explore the nutritional qualities of fish as food. Fish contains animal-quality protein. Thus, its amino acid balance and its content of essential amino acids is appropriate for consumption by animal species, including man. Plant proteins, on the other hand, generally do not have an amino acid composition appropriate for animals. However, with complementation, that is, the mixing of proteins from different sources, these proteins can be quite acceptable. Fish, either on its own or as a part of a protein mixture, is particularly high in the sulfur amino acids: cysteine, cystine, and methionine.

FATS AND OILS

Americans seeking to limit dietary fat often seek low-fat fish such as cod and flounder, which average less than 1% fat in the flesh. Twenty-five to thirty-five percent of the fatty acids of the fish fat are particularly long C20 and C22 fatty acids, that is, they are twenty and twenty-two carbon atoms long; the more unsaturated fatty acids with five and six double bonds may account for 15–30% of the total fatty acids. Fish are also high in the omega-3 family of fatty acids, that is, the linolenic family; and are lower in the omega-6 family of fatty acids, that is, the linoleic family. (In the omega-3 family of fatty acids the first double bond occurs after the third carbon from the methyl group, while in the omega-6 family it does not occur until after the sixth carbon.)

Recent research suggests that the omega-3 fatty acids affect the balance of prostaglandin production in humans in such a way as to slow down the blood clotting process. For Eskimos and others who eat great quantities of fish—or, really, seals—this might be a problem during an injury. But for many Americans, the slower clotting time seems to help prevent heart disease and strokes. The latter are blood clots that form in the brain, so that their prevention may yet lead to an official justification of the old wives' tale that fish is "brain" food. Sometimes, omega-3 fatty acids are consumed indirectly, for example, humans consume them when they eat poultry that have been fed diets containing fish meal.

The benefit of fish oils has been recognized by the Heart, Lung, and Blood Institute, a section of the National Institute of Health, and by the American Heart Association. These groups have promoted the consumption of fish in the United States. Unfortunately, encapsulated mackerel and other fish oils are now marketed in health-food stores; the prices are very high and the benefits of this approach, if any, have not been determined, especially since many consumers take these pills in *addition* to other oils they are eating rather than *instead* of these oils. None of the major health organizations, public or nonprofit, have come out in favor of fish oil pill consumption.

There have been other nutritional experiments done with fish as food. Marian Childs, University of Washington, suggests that the noncholesterol sterols in shellfish might actually inhibit the absorption of cholesterol by humans, thereby leading to a reduction in serum cholesterol levels. These oils may also be beneficial in the treatment of arthritis or other body-joint inflammations. For example, the green-lipped mussel (*Perna canaliculus*) has been shown to provide relief for arthritis sufferers. These findings are still new, and much research remains to be done. Nevertheless, we know that the specific essential fatty acids usually associated with all of these beneficial effects are eicosapentanoic acid (EPA, C20:5) and docosahexaenoic acid (DHA, C22:6). We can certainly expect more research on EPA and DHA in the coming years.

The NMFS Charleston laboratory has been given the mandate to produce various fatty acids for scientific purposes. They are currently using a selective precipitation system involving urea/ethanol. A list of omega-3 fatty acid quantities in selected finfish is shown in Table 14.1.

The presence of a high level of unsaturated fat in fish raises the potential for fat rancidity. Hematin, a breakdown product of hemoglobin and myoglobin in fish, is known to be a pro-oxidant. It is surprising that the rate of fat oxidation in fish is as slow as it is. We must therefore presume that some amount of antioxidant is naturally present in these animals. Vitamin E is one of many compounds being suggested as playing this natural antioxidant role.

The measurement of the rate and/or amount of fat rancidity in most meats is a difficult analytical problem and fish is no exception. The usual procedure is to measure thiobarbituric acid or TBA values and/or the measurement of peroxide values. Both of these values may peak and then drop again.

Table 14.1   Omega-3 Fatty Acid Content of Some Finfish.

Omega-3 Fatty Acid per 100 g Raw Fillet[a]

0.5 g and Under		0.6–1.0 g		More than 1.0 g	
Sole	0.1	Channel catfish	0.6	Rainbow trout	1.1
Northern pike	0.1	Red snapper	0.6	Cisco	1.1
Pacific cod	0.1	Yellowfin tuna	0.6	Pacific mackerel	1.1
Atlantic cod	0.2	Turbot	0.6	Atlantic herring	1.2
Walleye	0.2	Thread herring	0.6	Pacific herring	1.2
Yellow perch	0.2	Chum salmon	0.6	Sardine	1.2
Haddock	0.2	Striped bass	0.7	American eel	1.2
Yellowtail	0.2	Wolffish	0.7	Atlantic halibut	1.3
Sturgeon	0.2	Spot	0.8	Sablefish	1.3
Rockfish	0.3	Swordfish	0.9	Atlantic salmon	1.4
Brook trout	0.3			Lake trout	1.4
Silver hake	0.4			Anchovy	1.4
Striped mullet	0.4			Coho salmon	1.5
Atlantic pollock	0.4			Pink salmon	1.5
Ocean perch	0.4			Bluefin tuna	1.5
Carp	0.5			Atlantic mackerel	1.9
Pacific halibut	0.5			King salmon	1.9
Pacific whiting	0.5			Spiny dogfish	1.9
Weakfish	0.5			Albacore tuna	2.1
Skipjack tuna	0.5			Sockeye salmon	2.7

From Nettleton (1985).
[a] Figures have been rounded to the nearest 100 mg. Data represent the sum of eicosapentaenoic acid (EPA) and docosahexaenoic acid (DHA).

Another problem is to determine the relationship between the absolute value of the measured parameter, that is, the TBA number or peroxide value, and the organoleptic perception of the consumers. Even trained panels do not generally classify sample fish products as rancid at the same TBA or peroxide values; such values would probably vary even more for consumers.

The anatomical distribution of fish fat varies in different species. In low-fat fish like cod, the fat is mainly deposited in the liver. The liver may be 10% of the fish's body weight and contain 80% of the fish's body oil. This fish oil is particularly high in vitamins A and D—which is why fish oils were so popular with parents, but not children, for so many years. Unfortunately for the fishing industry, and possibly parents, synthetic vitamins have at least temporarily taken over this market. Kids, however, are not complaining.

High-fat fish have as much as 20% fat, usually found in different tissues. For example, salmon has a fairly small liver: it is only 1–3% of the total salmon weight and it contains less than 5% of the fish's total oil. In herring, the bulk of the fish's vitamin D is found in the flesh, which is high in oil. Sharks vary with respect to the amount of fat they have; most species have less than 2% fat, but some may have up to 13% fat.

The amount of cholesterol in seafood has recently been reinvestigated with more modern techniques that eliminate the bias due to the presence of plant sterols. With these new techniques, most of the supposedly high-cholesterol seafood have been placed at the lower levels common to other flesh foods. It is now believed that about 10% of crustacean and 60–70% of molluscan sterols are noncholesterol. Shellfish, although low in fat, have about 0.2–1.0 g omega-3 per 100 g of edible tissue. Crustaceans and molluscs are about 35–45% omega-3 as a percent of total fatty acid. The noncholesterol sterols, like those of plants, may interfere with the absorption of cholesterol.

VITAMINS AND MINERALS

Finfish have been shown to contain the minerals phosphorous, potassium, iron, copper, zinc, magnesium, selenium, cobalt, iodine, and fluorine. The high iodine content of marine fish is important for the prevention of human goiter problems. Oysters are the best source of zinc of any common food. As we move to improve the calcium content of the United States diet, the need to maintain the appropriate zinc levels in the diet may become more critical because of competitive binding on some enzymes between calcium and zinc, where the active form of the enzyme needs to be a zinc metalloprotein.

It is important to deal with selenium carefully because the necessary dietary level may be only 2 to 5 times lower than the safe level. If the normal "100 times" safety level used with most food additives were applied to the control of this compound, we would all be selenium-deficient. Selenium is now becoming a "magic" mineral. Appropriate levels of selenium in the diet are currently being credited with almost miraculous properties: reduction in cataracts, decrease in liver diseases, prevention of aging, prevention of cancer, reduction in cardiovascular and muscular diseases. If even a few of these benefits turn out to be real, it is going to challenge nutritionists' skills to develop safe ways of delivering a specific dose of selenium to the general population. There is no scope for megadoses here!

The calcium content of fish bones, particularly canned, is readily obtainable because the bones are very digestible. The retort process softens the bones so that they are easily pulverized. The iron in the marrow, heme iron, is of particular benefit to premenopausal women. Heme iron is more digestible than nonheme iron. About 10% of the heme iron is absorbed, while only 5% of the nonheme iron is absorbed. The presence of heme iron in the diet also increases the absorption of nonheme iron.

Table 14.2 Calcium Content of Some High-Calcium Foods.

Food	Quantity	Calcium Content
Collard greens	1 cup	304 mg
Nonfat dry milk	1 cup	375 mg
Sardines with bones	3 oz	372 mg
Yogurt, low-fat	1 cup	415 mg
Salmon, canned	3 oz	285 mg
Cottage cheese, low-fat	1 cup	138 mg

Courtesy of Maine Sardine Council.

Calcium levels listed on product labels support the contention that eating canned fish with bones is a good source of calcium. Some comparative levels are seen in Table 14.2.

Notice that all of the nonfish products are 1 cup, which is 8 oz of liquid measure, compared with 3 oz of fish flesh and bones.

Roe is particularly high in vitamins B^1, B^2, and B^{12}. As mentioned earlier much more could be done with this valuable resource.

Canned salmon—but not canned tuna—seems to show a thiamine reduction during canning. The niacin content of tuna and salmon may be higher than that of groundfish and may be associated with how actively the fish are feeding at the time of harvest. Another important consideration is the leaching of vitamins into the processing liquid and whether this is or is not included in the final nutritional value calculations.

Salt.   The relationship between high salt intake and hypertension has caused the American population and government to pay more attention to this ingredient. The whole issue of dietary salt as a cause of hypertension has recently been clouded by reports concerning the importance of higher levels of calcium in the diet to prevent hypertension development. The interactions between salt and calcium levels remain to be worked out.

Fish, including the marine fish species, are generally considered to be low in salt but not salt-free. The products that are high in salt are some of the processed products. Interestingly, the use of polyphosphate dips during fish processing may actually inhibit the uptake of salt. However, the polyphosphates may still contribute some sodium, if the sodium salts are used.

As part of its nutritional labeling efforts, the United States government has been trying to persuade more companies to label the salt content of their products voluntarily so that a mandatory salt labeling law will not be needed. The further processed fish industry finds it difficult to comply because many of its processes are not sufficiently well controlled; consequently, the products cannot be maintained at a consistent salt level. Legally, a company must insure that any product purchased by the consumer has less than 120% of the declared salt level. Many companies feel that they would have to put an extremely high figure on their label to cover the occasional sample that runs very high.

The United States government currently prohibits the advertising of medicinal properties of foods. Even placing a booklet, for example, advertising the benefits of fish oil medicinally, near a fish counter is currently illegal. However, the Federal Trade Commission, which regulates advertising, is considering a somewhat more liberal policy. It would be interesting to see whether such a change would change consumer habits. Note, for example that low-sodium tuna has not been successful in the marketplace, capturing less than a 1% share. The FDA is currently reexamining its policies in this area and at the time of this writing (1990), has written to all known manufacturers of fish oil pills asking them to remove any claims of medical benefit from the labels of their products.

Counting Calories.   The caloric content of fish can generally be calculated from its fat and protein content, that is, % fat × 9 representing 9 calories per gram, plus % protein × 4 representing 4 calories per gram. Table 14.3 is a table of reported calorie values. A full listing of the composition of fish and fish products, and many other products, can be obtained from the United States Department of Agriculture's Handbook No. 8. This handbook is now published in many separate sections in looseleaf form so that it can be updated regularly. Figure 14.1 includes some comparisons of fish products with other flesh foods.

These data are limited in their use. Using the aforementioned formula, the reported values for the calories usually agree with the composition data. In some cases, composition data and caloric value data may have been derived from different analytical results. Among the problems with all such tables is that the final data are only as good as the input data. With composition tables, the variability of the real product might be great enough to make it difficult to use in such a format. Note that the values in Table 14.3 were all taken from a single published listing! But given the broad range of composition expected for fish with respect to season, fishing grounds, and so on, the value of the data becomes extremely limited. In summary, although such data are desired by consumers, it is not clear what value they really have.

The contents of some vitamins and minerals in surimi is lower in comparison to other seafood, for example, those of niacin, calcium, iron, and sodium. Because of their importance to human nutrition and the significant potential differences between surimi and "real fish" products, the United States government has chosen to require that surimi seafood products be labeled as "imitation."

Table 14.3   Calories per 100 g of Various Fish

Species	Calories (per 100 g)	Protein (%)	Fat (%)	Sodium (mg)
Bluefish	107	19.2	3.3	68
Cod	74[a]	17.4	0.5	90
Croaker	92	18.6	2.0	80
Flounder	94	16.3	3.2	121
Haddock	77	18.2	0.4	98
Hake	82	17.0	1.4	
Herring	122[a]	17.7	2.8	105
Mackerel	167	19.5	9.9	94
Ocean perch	91	18.5	1.4	70
Shark	87	19.1	1.2	
Swordfish	118	19.4	4.4	
Tuna	145	24.7	5.1	100
Whiting	90	18.9	1.3	50
Carp	102	17.8	2.5	44
Catfish	157	18.2	8.2	60
Rainbow trout	154	20.7	6.8	52

[a] Calculated calories. (Note that some figures are correct on the basis of fat and protein content; others are not.)

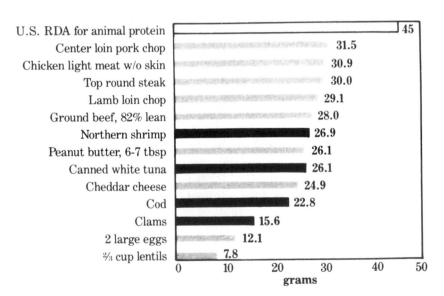

(a) Protein content (g) per 3½ oz cooked serving.

Figure 14.1   Comparison of selected nutrients for various flesh foods. (*Nettleton 1987, pp. 4, 6*)

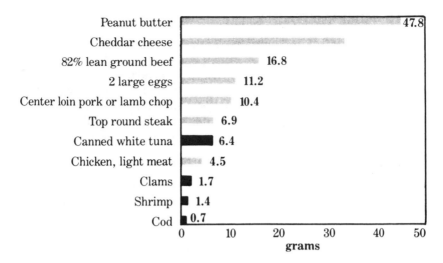

(b) Fat content (g) per 3½ oz cooked serving.

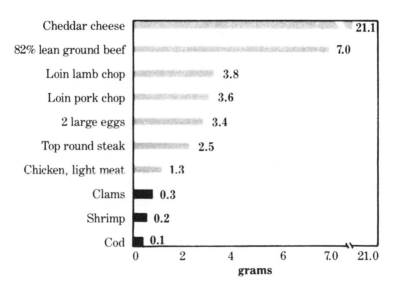

(c) Saturated fat content (g) per 3½ oz cooked serving.

Figure 14.1 (*continued*)

FISH CONSUMPTION

The per capita consumption of fish in the United States is about 15 lb per person per year. However, like many statistics on food consumption, this figure is misleading. How are such numbers calculated? They are supposed to represent fillet weights for finfish and edible parts for shellfish. In fact, the figure is estimated to be 35 lb of live weight product per person per year. Furthermore, these figures only represent commercial product, completely ignoring the significant contributions of hunters and fishermen (estimated at 4 lb/person).

Most per capita consumption data are obtained from wholesale trade movements and do not include losses that occur after the point of tracking, for example, with flesh commodities that are trimmed significantly following their departure from the slaughter house or fish that are further processed after they have left the processing plant. All of these figures are corrected for imports and exports. Their greatest use is for monitoring changes in production over time, but they are often misused in misguided attempts to determine the nutritional status of an entire population.

In 1989 an unusual occurrence took place. The initial announcement of the per capita consumption for that year indicated a jump from 15 lb in 1988 to over 17 pounds in 1989. Once the figures were released, it became obvious that something had gone wrong. On reexamining the figures, the NMFS statisticians discovered that the Alaskan pollock catch had been "calculated" on the basis of fillet yields rather than as "surimi." The correction for this single factor essentially lowered the value to 15.9 lb. Table 14.4 offers comparisons between reported values and the presumed amount actually consumed of flesh foods.

To evaluate the nutritional status of a population, the actual amount of food consumed must be measured more accurately. These values can only

Table 14.4   Reported Versus Actual Consumed Quantities (per capita per year) for Various Flesh Foods

Item	Reported Consumption (lb)	Actual Consumption (lb)	Reported/Actual (%)
Beef	120.5	43.8	36.4
Pork	61.4	25.1	40.9
Chicken	47.7	19.6	41.1
Seafood	13.4	9.7	72.0
Turkey	9.4	4.4	46.8
Veal	3.0	0.7	23.3
Lamb	1.6	0.5	31.3

Courtesy of Robert Hasiak.

Table 14.5 Fresh and Processed Meat Edible Weights

	Carcass[a]	Retail[b]	Fresh	Processed	Total
Beef	104	77	47	11	58
Pork	62	59	8	23	31
Lamb	4	3	2	0.3	2.3

[a] Edible portion, official per capita.
[b] Per capita yearly weight consumption.

be calculated if an accurate value of the weight of food consumed by a "typical" consumer in the designated population is obtained, which would include accounting for plate waste and other losses. The latter can be physically measured at the time of consumption or estimated by garbage studies, such as the famous Phoenix garbage studies.

It is also interesting to note that the percentage of red meats being consumed fresh, rather than processed, has been rising in the United States in the last year or two (Table 14.5). These figures are based on slaughterhouse production corrected for imports and exports.

In spite of the limitations of these figures, statistically oriented food scientists might find the following numbers interesting. It appears that the percentage of fresh finfish consumed in the United States is increasing, as compared to the total finfish consumption in the United States:

Year	Fresh/Total Finfish
1978	19.4%
1979	20.7
1980	23.4
1981	22.7
1982	25.3

PARASITES

The presence of parasites in fish flesh is of great concern to the fishing industry. Much time and effort is devoted to removing these parasites; marketing problems develop when consumers discover the occasional parasite in fish.

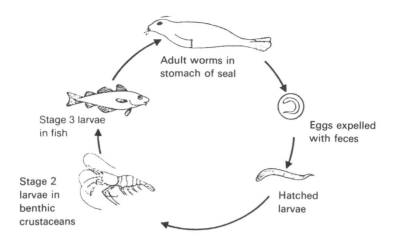

Figure 14.2 Life cycle of the seal worm. (*Faria, 1984, p. 81; Courtesy of Massachusetts Division of Marine Fisheries*)

The organism referred to as the "cod worm" is actually a nematode (roundworm) called *Porrocaecum* (*Terranova*) *decipiens*. Although it is sometimes found in the flesh of a live fish, it is more usually found in the belly lining. The worm does not generally move into the flesh until the fish's death. Brown to reddish brown, 9–58 mm long and 0.3–1.2 mm wide, the worms are not considered harmful to humans. Nevertheless, consumers do not want to find them—or eat them!

The alternate hosts for these nematodes are seals and other marine mammals. Thus as the number of seals increases, so does the number of worms in fish, particularly in cod. The feeding habits of cod and haddock are different enough that the level of worm infection is higher in cod than in haddock from the same area. It follows, then, that harvesting seals serves to decrease the infestation of marine parasites in cod. A summary of the life-cycle of the seal worm is shown in Figure 14.2.

One method of detecting seal worms is candling, that is, shining light through the fillet to make the worms visible. Operators then remove the worms by hand. It is difficult to develop an incentive system that rewards the operator for doing the best possible job in removing worms without slowing the line down too much.

A larger candling table makes candling easier. The light above the candling table should be about 35–50 foot-candles. One foot-candle is the amount of light a standard candle provides at one foot from the candle. The overhead light should be about 300–500 foot candles. Two lights should be included in the table to prevent 60-cycle flicker. Fresh bled fish are easier to candle than old fish. Large fish have more worms in the nape, small fish more in the loin; small fish have fewer worms altogether than large fish. Right/left fillet differences have been noted.

Other candling techniques are being explored. Ultrasound apparently detects the coiled worm if the orientation is correct. X-ray equipment is claimed to detect parasites, bones, and other foreign matter. Scanning laser acoustical microscopy detects the worms, but may not be commercially viable because of the need to examine the fillets in a waterbath.

Freshwater and anadromous fish carry worms that may be capable of implanting in humans if the worm-containing fish is eaten raw. In Jewish homes, these worms might be found in the "Grandmother" or whoever prepares the gefilte fish (see Chapter 13). Grandma tastes the mix before it is cooked, but everyone else in the family only gets to taste the gefilte fish after it has been boiled for an hour!

The culprit is usually the tapeworm, or flatworm, larva called *Diphyllobothrium latum*. The worm can grow up to 10 m (30 ft) long in the human intestines and causes a vitamin B^{12} deficiency. Although the disease can be cured with drugs, it is better to kill this organism and other freshwater parasites before ingestion by one of the following treatments:

54°C (133°F)	for 5 minutes
−18°C (0°F)	for 24 hours
−10°C (14°F)	for 72 hours

Marinating in strong brine (not in a mild pickling mixture) can also be used. Note in the above that at least 72 hours are needed in the home freezer.

Herring, salmon, and squid may also carry a colorless, round worm (*Anisakis simplex*). Smaller than the seal worm, this organism is killed by brining or freezing. Anisakiasis symptoms include cramping, abdominal pain, nausea, and vomiting. The organism responsible for milky hake is a protozoan called *Chloromyscium thyrsites*. Higashi (1985) has reviewed marine parasites in detail.

INSECTICIDES

In many underdeveloped countries where fish are routinely dried, the problem of insects becoming incorporated into the product is quite serious— particularly if the drying is done on the ground. In these cases, it has been estimated that over 20% of the protein in dried fish might actually be insect protein. The Tropical Products Institute of London recommends that pyrethrins and piperonyl butoxide be applied directly onto the fish. They also recommend fumigation efforts with phosphine to control these insects. Alternatively, the use of various indoor heating systems may bring the temperature high enough to kill the insects and beetles during processing.

BOTULISM

Botulism is a potentially fatal disease. *Clostridium botulinum* is the anaerobic toxin-forming organism that is responsible for botulism. Generally found on fish surfaces, *C. botulinum* is not particularly competitive, so if other organisms grow there, it generally will not. Types A, B, and F are proteolytic, so that growth is accompanied by a detectable off-odor.

Type E, the type most often found with fish, is nonproteolytic and has a minimum growth temperature of 3.7°C (38.7°F). It is found in various natural waters and can therefore be found with specific fish. At times it has become a problem in farmed fish, where the organism takes hold and spreads. Apparently, it does not affect the fish themselves.

There is some controversy over how useful the official methods of testing for botulism are in evaluating processes that might be used by the fishing industry. The spores are generally added in large numbers either onto the surface of the fish or deep into the test material via inoculation. The former is more appropriate for research related to shelf-life extension whose treatments deal with the fish's exterior. The test levels are 1,000 to 10,000 times above those ever found in fish. More traditional safety tests use a factor of 100. The highest level of spores ever found on fish was 20 spores/100 g of fish. But in some samplings, botulism spores were found on 25% of fillets. The results of the testing depend on the test level used: with high levels of spores, the chemical used to limit their growth may be "swamped out" by the vegetative botulism cells.

The method of calculating doses must be considered. The entire dose of botulism spores for the sample per 100 grams of fish may be injected into a single spot, thereby exceeding locally even the calculated safety factor. It is not clear whether this problem can occur during surface applications of spores.

The toxin produced by the botulism organism is normally destroyed by cooking, either as part of the canning process or as done at home by the consumer, so that the problem should only exist for raw fresh fish products. These are, however, popular dishes in many countries, for example, sashimi in Japan, ceviche in Mexico, and a lemon and raw coconut marinade in New Zealand.

There is a more legitimate concern with smoked fish where vacuum packaging may give rise to conditions of temperature abuse. The botulism organisms in these products may find themselves in an ideal situation: an anaerobic environment in the absence of any natural competitors. Also, the final product is often consumed without any further cooking, so the toxin itself, if present, would not be destroyed. A water phase salt content of 3.7% has long been considered the minimum level necessary to prevent botulism outgrowth. However, recent research by M. Eklund at the Seattle NMFS laboratory indicates that liquid smoke may be antibotulinic and may therefore permit lowering the salt level to 2.0%, a boon to modern health-conscious smoked fish consumers.

The most serious problems have generally occurred with processed foods produced at home by consumers. Sanitation and knowledge of the potential dangers may be lacking in many cases.

Nitrites and nitrates are antibotulinic compounds used extensively for this purpose with red meats and poultry. The use of nitrates has been severely limited in recent years, being restricted mainly to dry-cured hams. However, because fish has more nonprotein nitrogen, nitrite is more likely to form various nitrosoamine compounds when used in fish. Since many of the nitrosoamine compounds have been found to be carcinogenic, some people feel that nitrite should not generally be used with fish. Others, particularly regulatory agencies, specifically encourage its use, particularly with smoked fish, by permitting lower water phase salt levels to be used with products containing nitrites.

Preventing botulism outgrowth in fish and fish products demands the concern of everyone in the industry. However, it should not be waved as a red flag impeding legitimate and safe improvements in the technology for handling and marketing fish. The development of more "sous vide" products, that is, products pre-cooked in a vacuum bag, should give the regulatory agencies more experience with complex products that are protected from botulism by a series of incomplete multiple barriers, rather than a single factor.

HISTAMINES

Histamines are produced by enzymes that break down the amino acid histidine. Whether there is an endogenous enzyme in fish is unclear. However, the bacterial enzyme, that is, the histidine decarboxylase, has been isolated from a few different organisms. The temperature minimum for histamine production remains controversial: some species do not seem to be able to produce it below 10°C (50°F) while a few species do. *Klebsiella pneumoniae* was able to produce histamine in small amounts down to 4°C (39°F). However, the enzyme is not widely distributed among bacteria and is only found in certain enterobacteriaceae, clostridia and lactobacilli. Actual relationships of histamine poisoning to bacterial population center almost exclusively on *Morganella morgani* (previously *Proteus morgani*), *Klebsiella pneumoniae*, and *Hafnia alvei*.

A histamine level of 10 mg/100 g is considered to be significant, while 100 mg/100 g is considered to be toxic. The United States regulatory action levels in tuna state that 20 mg/100 g is a defect action limit and 50 mg/100 g is a hazard. (The unit of mg/100 g may also be expressed as mg%.)

The role of potentiators in the development of histamine poisoning needs to be better understood. For example, 150 mg/kg body weight of either histamine or cadaverine is harmless. Taken together they are lethal. Thus the role of histamine versus that of other biogenic amines needs to be more carefully considered. Cadaverine is produced from lysine, putrescine from ornithine. The bacteria that do this on/in fish are not currently known, nor are their temperature requirements.

SHELLFISH POISONS

Viruses that grow in shellfish can lead to gastrointestinal difficulties in consumers. The greatest cause of this problem is purchasing of shellfish from nonreputable dealers. The Interstate Shellfish Sanitation Conference continues its work on development methods to better insure the safety of shellfish. It is not clear whether cooking and depuration sufficiently remove these viruses. Fecal coliforms are not necessarily a good indicator of their presence.

A number of other organisms can become associated with shellfish and cause public health problems when the shellfish are consumed. Paralytic shellfish poisoning (PSP) is caued by *Gonyaulax tamarensis*, a dinoflagellate associated with red tide. Because the red tide is usually visible in the water, it is easy to identify when the problem is occurring. Shellfish harvesting from these areas is then prohibited. It would seem that the visibility of red tide to the harvester and the regulatory agencies would help simplify the enforcement problem. However, there have been reports of outbreaks when the visible red tide was not a predictor of problems. Worse, there have been "red tide" problems without sufficiently visible red tide bloom to serve as a warning. *Gonyaulax tamarensis* has not been found in oysters or in the edible portion of scallops, but can be found in great quantities in the scallop viscera. Part of the problem is that the organism apparently overwinters as a cyst. Animals that eat these cysts become mildly toxic. The toxin is about fifty times more toxic than curare. PSP leads to symptoms such as dizziness, shortness of breath, nausea, and a loss of sensation at the extremities. The current testing method involves a mouse assay. Various analytical methods have been proposed to detect "saxitoxin" and related neurotoxins. The FDA is currently testing a high-pressure liquid chromatographic method (HPLC). Kits using ELISA (immunological) technology are also being developed.

Note that neurotoxic shellfish poisoning is also caused by a dinoflagellate.

Diarrhetic shellfish poisoning is caused by a lipid-soluble toxin which causes diarrhea. Symptoms usually appear within 12 hours and may last for up to three days. The offending compounds are believed to be polycyclic ethers. The shellfish can be rendered toxic with as few as 200 of these cells

(Courtesy of E.C. Publications, Inc.)

per liter of seawater. In the United States and Europe the causative organism is *Dinophysis acuminata*; in Japan and New Zealand it is *Dinophysis fortii*. Other *Dinophysis* and *Prorocentrum* species of dinoflagellates have also been identified as carrying the toxin.

CIGUATERA POISONING IN FINFISH

Ciguatera poisoning has been traced to the dinoflagellate *Gonyaulax toxicus*; however, other dinoflagellates are also suspected. Worldwide incidence is about 50,000 cases per year, usually caused by tropical fish. Symptoms include gastrointestinal inflammation, cardiovascular irregularities, and neurological symptoms such as tingling and numbness of the extremities. These symptoms can last for months. No known cure exists. (Protamide has been used in other countries with some success. It is not approved for use in the United States.)

The problem has got worse with new pressures to use different fish for food. Fishermen are tempted to fish in new areas or buy fish from distant fleets. There is no way to detect the problem, so control is difficult. A radioimmunoassay exists, but is not a practical test to run regularly. Unfortunately, random sampling is not enough because only a few fish may be toxic within a given catch. Better methods of dealing with this problem must be found if more extensive utilization of tropical fish is anticipated.

REFERENCES

Faria, Susan M. 1984. *The Northeast Seafood Book*. Boston: Massachusetts Division of Marine Fisheries.

Higashi, Gene I. 1985. Foodborne parasites transmitted to man from fish and other aquatic foods. *Food Technol.* **39**(3): 69–74, 111.

Lands, William E. M. 1986. *Fish and Human Health*. Orlando, FL: Academic Press.

Nettleton, Joyce. 1985. *Seafood Nutrition: Facts, Issues and Marketing of Nutrition in Fish and Shellfish*. Huntington, NY: Osprey Seafood Handbooks.

Nettleton, Joyce. 1987. *Seafood and Health*. Huntington, NY: Osprey Books.

15

Marketing

The proper marketing of seafood is an important aspect of improving the economic viability of seafood products. The fishing industry used to be "commodity driven," that is, selling (often in the form of "unloading" or "getting rid of") today's catch was the dominant theme. It is now trying to be an industry that markets and promotes its commodities in a more sophisticated fashion. If fish is perceived as a quality product, the industry could see substantial growth.

The goals of this chapter are to develop a vocabulary base for members of the fishing industry and suggest areas of special market considerations. Since this is not the authors' primary area of expertise, we will recommend books for further study.

PRICING

By marketing, we mean the selling of a product using a full range of promotional methods in a way that increases product usage and increases industrial profitability. With effective marketing, a product can become sufficiently differentiated that its price is not solely determined by total supply, that is, it will have created its own independent consumer demand.

Classical economics uses supply and demand to determine the price of a product. Supply represents how much product is offered for sale. This normally refers to a specific product of interest, but how specifically should the product be defined in the fishing industry? Do we refer to all cod, cod of the size required for a certain process, only fresh cod of a given size? Formal economic analysis usually focuses on narrower units, for example, fresh scrod cod; but in the "real marketplace," other components of the cod supply will affect buyers' perception of the available supply.

Demand represents consumers' potential purchasing power. Once again, a narrower view focuses on the demand for a particular product. But demand in the real world encompasses a wider perception. Some of this wider perception can be formally measured using the price elasticity concept, a measure of how much a product's demand changes as its price changes. If another product can easily be substituted for the original product, then the demand is price-elastic, that is, a small change in price shifts demand to the other product. On the other hand, a unique product in the market may be less price-elastic, that is, even a change in price does not change demand significantly. A particular elasticity may only hold over a certain price range; if price goes too high, substitutions occur and demand decreases, tending to drive the price back down. If the price goes too low, then new buyers may enter the market, creating increased demand that may drive the price back up.

All of this theory refers to a "free" market with many buyers and sellers. For any given product, it is important that enough buyers and sellers participate and that they have equal access to pricing information, that is, that there is a respected price discovery method, such as an "auction" structure. However, if many private transactions are reported publicly, the necessary information can still be obtained by all parties. Many people in the United States fishing industry are concerned that too many transactions take place privately, thereby jeopardizing the price discovery mechanism. Commercial companies, such as Urner Barry, will call a number of companies every day and learn about their transactions so that they can compile a reasonable price listing with sufficient information to give a pretty good picture of what is happening in the market.

The price of a product with a longer shelf-life, for example, frozen product, is affected by the amount of product being held in cold-store warehouses at a given time. Industry members familiar with the market monitor changes in warehouse "holdings," and concomitant seasonal adjustments, and get a good idea of the overall supply/demand balance. Small changes in these holdings can sometimes have a dramatic effect on prices as buyers and sellers try to adjust their long-term commitments.

A commodity speculator is someone who picks up some of the financial risks being carried by buyers and sellers who actually need the product. In return, the speculator may make or lose a lot of money without having a direct connection with the product itself. Such a system requires a product whose description is specific enough that there is no debate about what product is being bought and sold. Unfortunately, the fishing industry has not yet reached this degree of sophistication.

Prices can be established by other means. The Fisheries Cooperative Marketing Act of 1934 permits fishermen to act in association in collectively "catching, producing, preparing for market, processing, handling and marketing" their fish. It is also possible for them to consolidate their selling power. Such co-ops give fishermen the right to activities that would be illegal for commercial companies, such as agreeing not to sell a product unless they obtain a certain price. The idea is that an individual fisherman would be at the mercy of much stronger incorporated business organizations; this legal support is meant to create a more level playing field.

It should not be surprising that the market for fresh fish is usually more limited and regional; the market for further processed products and alternate shelf-stable forms of fish such as canned, pickled, and dried are much more international. In the latter case, transportation costs, reputation of the dealer or the country of manufacture, and quality standards will affect the price of the product. For example, for many years frozen Canadian cod was discounted on the world market as compared to Icelandic cod, sometimes as much as 40 cents on a product selling for under two dollars. As Canada has improved the quality of its product, the sellers have been able to increase the price to create more direct competition. The lower quality of Canadian cod gave the entire industry a bad name. Thus, the other countries involved in this market, particularly the Icelanders, were happy to see Canadian quality improve and for Canada to get full price for their product. There is an important lesson here—everyone in the industry "for the long haul" suffers when the consumer is given a poor-quality product.

SELLING PRODUCTS NATIONWIDE

How do fishermen or a fishing company actually sell products? Sometimes they do it themselves, either as individuals or through a sales department. This may involve some formal marketing mechanism, but often it just means a conversation between the buyer and the seller, who negotiate until they arrive at a price acceptable to both parties. Sometimes a company uses a third party to assist in this process. If the third party actually takes possession of the products—and is thus both buying and selling—they become the traditional middleman. For example, a distributor in an inland city obtains fish from various fish processors, "fish houses" on the coast, to fill his or her orders. The distributor then assembles the orders and delivers them to the various local units.

In other cases, a company, often a foreign company, can use a broker. Brokers work for the fish company and receive a fee, that is, a commission, for their effort. The difference between brokers and company salespeople is that brokers, or brokerage firms, only earn money if they sell the product. A broker can take on many different accounts and collect a percentage of sales from each of the accounts serviced. A company salesperson can only sell the company's fish and hope the supply does not run out. A brokerage firm can bring other advantages. By covering many different products, the firm's sales personnel can sell a wider variety of products to a single customer, thereby reducing the cost per sale for a company with a limited product line.

Also, by hiring brokers throughout the country, a company can widen its sales area without having to maintain and manage staff around the country. The fishing company can concentrate its efforts on producing high-quality fish products.

SELLING PRODUCTS INTERNATIONALLY

Selling internationally is always complex because special procedures are necessary whenever a product has to cross a border, and these procedures vary somewhat from border to border. New participants become involved in the process. Freight forwarders are experts in laws relating to shipping products and companies available to do the work. Like a broker, freight forwarders can do the work of an in-house shipping department, often at a cheaper, more efficient price for smaller companies. Custom brokers specialize in laws concerning duties and tariffs on goods that are going through customs formalities and payments. Many American fish companies have been hesitant to ship internationally because of the additional effort involved, but those who have done so often find that it can be a very successful and manageable business.

International pricing information is also more difficult to obtain. Various services are available to help provide up-to-date fisheries information, including movement and pricing information. One example is Globefish, an international project of the United Nations Food and Agriculture Organization.

Actual payment is often handled using a letter of credit, often an irrevokable one. This means that the buyer—usually through a bank in the buyer's country—arranges for a letter of credit for his or her bank to be sent to the seller. At the appropriate time, the seller can take the letter to his or her bank and receive payment. For example, when the product clears customs in the buyer's country, the letter of credit becomes valid. This is a simplified description of the process; since all of these dealings are complicated, professional assistance is essential.

Several United States fisheries development councils have published booklets and manuals to assist the reluctant fishing industry in overseas transactions. It takes some effort to iron out the system's bugs, but international trade presents many new opportunities. These development councils often have Telex and facsimile facilities that fishing industry staff can use at little or no charge.

PROMOTING PRODUCTS INTERNATIONALLY

How does one find out about potential buyers in other countries? Possibilities include ads in appropriate magazines and newspapers, listings with the commercial section of US embassies in those countries, and salesperson/broker arrangements in the target market nation. Another mechanism for selling on the international market—or for that matter, the domestic market—is the use of trade shows. A lot of selling and buying takes place in this forum, sometimes at the official vendor booths, often simply because buyers and sellers have a place to meet. Trade shows are an excellent place to learn about the marketplace for many products.

The United States government subsidizes efforts at overseas marketing including fish products, particularly at trade shows where a "United States delegation" may be assembled and the government may arrange for a country pavillion or booth. In other cases, an individual state government might take a booth at an appropriate trade show; all the companies from that state can exhibit at a significantly lower cost than if each had its own booth. The National Marine Fisheries Service seems to have discontinued these important programs, although some assistance from the Department of Agriculture is still available.

It has to be clear what is being sold. Is it a straightforward product such as a fish fillet of a traditional species? Or is it a fish fillet of a species with which the market is less familiar? In the latter case, potential customers need to see it and taste it. Sampling becomes extremely important and a great deal of effort should be put into insuring that appropriate samples are available during the trade show.

It may be that a company has developed a completely brand new product, particularly a further processed product that is no longer sold from the fish counter. The future of the fish industry may well depend on much more products being offered to both consumers and institutional markets. Extensive product development work must be done before the product can be shown. The marketing concept must be defined and backed up by appropriate feasibility studies in both the technical and the business domains. Food product developers and marketers must work with food technologists to understand the whole world of technical materials, such as commercial ingredients, available in formulating a product. But there are also engineering questions that must be answered to make mass production, packaging, and distribution possible. And food safety concerns must also be addressed. All of these aspects must be included when calculating the actual production costs.

Throughout the process, the entrepreneur must keep the team's work in touch with consumers' needs. And of course, even with the best of products, the world does not beat a path to one's door; one must have an excellent marketing plan and then execute it well. Is it any surprise that fewer than 10% of the new food products introduced into the marketplace are successful? For information on positioning a new product in the marketplace, we recommend the book of Ries and Trout (1986b). Part of the process is knowing how to get a project done. Daniel Best of *Processed Prepared Foods* suggests "10 ways to kill a project":

1. Overlooking key product attributes. Rarely do products base their appeal on a single attribute. "Total concept engineering" is the process of defining and then using engineering to include all the desired attributes.
2. Overlooking the resource network. Do not forget that the suppliers can be a great help.
3. Reducing quality to reduce costs. This can constantly chip away at a product.
4. Misapplication of resources. Give the required time, labor, and capital to the project.
5. Misuse of consumer test tesults. Remember that these tests only distinguish shades of gray.
6. Taking costs at face value. If the product looks as though it will be too expensive, check all assumptions and projections carefully before considering rejecting the project.
7. Lack of commitment.
8. Overlooking the time factor.
9. Tactics versus strategy. The tactics of the short-term must support the long-term strategy of the product.
10. Management by avoidance. Do not depend too heavily on others.

TARGET MARKETS

Supermarkets and Other Retail Outlets

Supermarkets and other retail outlets serve individual customers. Store size varies over a wide range and greatly affects the number of items stocked. The largest supermarkets can run up to about 15,000 SKUs or "individual stocking units." (Three different sizes of the same product are counted as three different SKUs.) The typical supermarket stocks 18 different fresh and 59 different frozen fish and seafood items. Fresh product accounts for 42% of sales and frozen for 57%.

Customers normally select items and place them into shopping carts without interaction with store personnel. This contrasts with small retail stores in which there is often significant interaction between customers and store personnel. In recent years, large supermarkets have initiated the small retail store model for certain items, for example, providing full service fresh fish counters.

Retail stores generally receive most of the products they stock from a central warehouse distribution point; a few items, such as bread and milk are delivered directly to the market by the manufacturer/distributor, and are referred to as "store-door" deliveries. Fish can be delivered either way, but the former method may add a day or two to the distribution scheme. Kroger, for example, uses a single warehouse in the United States midwest to service all of its 1,600 stores, some of which are in Florida, that is, over a day away by truck. At the other extreme, Cub Supermarkets in Minneapolis uses an overnight delivery service to deliver fish quickly from warehouses on each coast to every single operating unit. Delivery to a warehouse may be more efficient—one stop versus many stops—but there is a potential loss of product quality.

The new product manager will be introduced to other costs in this market. Although not yet common for commodity items like fresh and frozen fish, supermarkets are currently charging "slotting allowances" and other—sometimes significant—fees for handling any new item through their warehouse system. Supermarket chains also expect both new and continuing products to "support" the supermarket with assistance in promotionals, such as participation in store ad programs and providing in-store demonstrations, coupons, point-of-purchase material, and free-standing advertising. If the product fails, some supermarkets charge a "removal" fee. More often, commodity items are requested to offer support in programs like the Nutrifact program, which provides basic nutritional information at the point of sale. Support is an important consideration in a supermarket's marketing strategy when it is approached to sell a new product.

Another aspect that sometimes surprises the newcomer to this market is sensitivity to payment terms. That is, when does the supermarket have to pay the supplier? Is it 10 days' net or 30 days' net? The supplier has to be able to cover the cash flow until payment is made.

The supermarket may already operate a pre-packaged fish program, a modified fish service operation, or a full service operation. This influences the type(s) of products the fish section can handle and the types of customers the store attracts. Full service often helps to sell a product, but it makes certain demands on a store that functions predominantly in self-service mode. In the self-service section, the supermarket's major challenge is to have the correct mix of products available to the customer at appropriate prices; in a full-service section, the appearance of the operation and the quality of the personnel become additional important concerns.

The small retailer, that is, the fish shop, is generally a full-service operation that is run by someone knowledgeable about fish. Smaller shops are not as efficient as supermarkets and, generally, have higher prices; but they are also perceived as offering a higher-quality product. This may be why smaller specialty shops are enjoying renewed popularity, especially in upscale markets. In urban areas, particularly in an ethnic neighborhood where the population are fish eaters, small stores can operate even in less-affluent neighborhoods; this is particularly true if the tradition of small-store shopping still thrives. The operators of these stores often obtain fish directly from coastal purveyors and pay slightly higher prices for shipping and handling relatively small orders. Others obtain their fish through distributors. Still others use the city's wholesale market, for example, the Fulton Fish Market in New York City.

Institutional Markets

Institutional markets are varied. There are large centralized operations such as hospitals, prisons, and schools. At the other end of the spectrum, there are individualized food service operations such as fancy restaurants that serve very few tables a night. The larger units often buy food products on bid, that is, a formal product specification is defined, along with a procedure for companies to qualify to bid. After the user reviews the bids, one company is awarded a contract to supply the agreed product(s), usually for three months to a year. The deliverable is often frozen or otherwise further processed so that distribution can be handled more easily. Institutions generally require government inspection and certification, that is, participation in the NMFS voluntary seafood inspection plan.

Other sectors of the food service industry, such as the fast food stores, buy mostly frozen products; larger restaurant chains generally use a mixture of fresh and frozen. In most cases, purchasing is done through the institution's central buying operation. This means that the supplier must be able to supply a great deal of product during the contract period. If the contract is not renewed, this can have a significant impact on the supplier's business. In a recent example, fast food operators are interested in adding catfish to their menu, but very few members of the fishing industry can guarantee sufficient supply. On the other hand, when Church's, a fried chicken chain, added catfish to its menu, the early sales were not as good as the company had predicted in spite of extensive promotion. Had the project been successful, it would have represented about 25–30% of all the catfish produced in the United States; but instead the project had to be dropped. It is interesting to monitor the progress of two similar projects and their impact on the raw material cost and supply; shrimp salad sold at McDonalds and Taco Bell.

Smaller restaurant accounts require a great deal of service: high standards must be maintained for small orders. However, these accounts offer a real potential for the local fisherman or aquaculturist with an interest in "niche" marketing. The restauranteur and fisherman must work together to overcome standard problems. For instance, fresh product can only be offered for a limited time period, that is, when product is available; but most menus cannot be reprinted more than once every three months or so. With good communications, however, the restaurant can use daily "special" placards in their menus or menu boards that reflect what is available in the marketplace.

MERCHANDISING MODEL: FROM THE SUPERMARKET TO THE HOME

The "Wet Fish" Case

The size of the display and the product mix are generally worked out according to local consumer demands. Because consumers often purchase on impulse, it is important that the display attracts customers to the area and helps them focus on those fish that the store wishes to promote. The display should include greens, lemons and the like to add some color; the distribution of product should offer some visual contrast. There are a number of booklets and videotapes available to aid supermaket staff in setting up a visually attractive, safe display case.

Personnel must use sufficient ice to provide the necessary cooling, and must occasionally mist the display to prevent dehydration. If the fish are stacked too high, then some of the product is too far from the cold source. On the other hand, fillets should never be in direct contact with the ice. Fresher product should be towards the customer, older product towards the back where store personnel can sell it using a "first in, first out" policy. It is important to label products carefully with the date they are received so that first in, first out can also be maintained in the product storage cooler and in the central warehouse.

Figures 15.1 and 15.2 illustrate the wrong and the right way to set up such displays. In recent years a new philosophy has developed. Ice is used simply as a garnish and the display case is maintained at $-2°C$ to $0°C$ ($29°F$ to $32°F$) by mechanical refrigeration.

Cooked and raw products must be kept apart. Personnel handling product must not touch cooked product directly after handling raw product; food handlers should not handle money intermittently. If this is not possible, the seller should at least try to wash his or her hands before handling cooked, ready-to-eat products. The price stickers do not belong directly on the fish. Attention to general sanitation of the facility is also very important.

Enhancements for the Customer

Packaging.   Chapter 5 deals with packaging for retail.

Positioning in the wet case.   Market-related products such as sauces, batters,

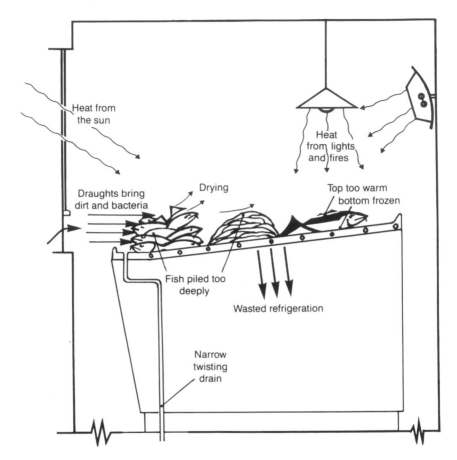

Figure 15.1 The wrong way to maintain a wet-fish display case. (*Aitken et al. 1982, p. 54; Crown copyright*)

and canned soups in the store's seafood section can improve the area's profitability. Promotions can be built around interesting combinations of products that obviously enhance each other.

In setting up the wet fish display and these related products, personnel must be aware of "positioning," that is, the actual placement of products and the number of "facings," that is, how many rows of a particular product are seen by the customer of each item. Items that are too high up or too low down are more poorly positioned; items at the end of the fish counter or near an aisle end are better positioned than items lost in the middle of the display. The ends beyond the aisle are considered particularly prestigious; more facings suggest a more important product. Burying fish products in

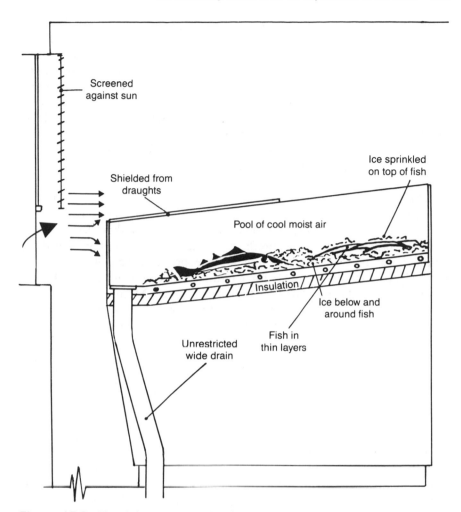

Figure 15.2 The right way to maintain a wet-fish display case. (*Aitken et al. 1982, p. 53; Crown copyright*)

the meat department is disastrous, especially when the meat manager does not know (or care to know) how to handle them.

Coupons. The use of discount coupons can be an important way of promoting a product in the United States. It is used both to encourage customers to try new products and as a way to develop brand loyalty. The most important source of coupons for most consumers at this time is the free-standing insert in the Sunday newspaper. These inserts appear on Saturdays in small towns without a Sunday paper.

However, more important than the percentage of coupons distributed in different ways is the rate of redemption of the coupons found in different media, because this is the real measure of the impact of those media in the marketplace and of the potential liability undertaken, that is, how much financial exposure the company has with a particular coupon program. Coupon redemption rates are quite low, ranging from an unsatisfactory 1 or 2% up to a highly successful 7 or 8%. Professional assistance is available and probably necessary to achieve the higher redemption rates.

Another trend that is developing is a higher value for coupons. The higher the value, the better the redemption rate.

Branding.   Branding is the labeling of products with the name of the specific processing company. The poultry industry took chicken from a commodity item, often sold solely on price, to a branded product that the customer selects by brand and that the specific manufacturer promotes; in return the manufacturer may earn a premium for the product. The processor of a branded product is responsible for handling any complaints, even if most complaints are first presented to the supermarket. Monitoring consumer complaints well is obviously important so that repeat customers are not lost. But more importantly, the process creates insights into customers' needs for future marketing.

Branding has had a very positive impact on the poultry industry. Is the fish industry ready for this? To brand, one needs a consistent product of high quality. In principle aquacultured product like catfish and trout are often branded; but in practice the lack of serious retail advertising, to make them recognized "household" brands, qualifies the "branded" definition. The advertising is mainly aimed at the trade. There are other ways to distinguish different brands of a product from each other. The use of "store brands" gives most of the recognition to the supermarket itself. If the program is run well, the consumer has more incentive to purchase many products at that store. These "private label brands" are generally used by supermarkets as part of a unified theme throughout the store. To date, however, fresh products like fish have not been included in supermarkets' private labeling efforts.

The Supermarket Environment.   Many supermarkets now use automatic scanning processes to speed check-out. This procedure requires that each checked-out product has a universal product code (UPC) symbol printed on its package. A slightly different coding system is needed for "random weight" products such as fresh fish. A UPC coding system for seafood products has been developed by the National Fisheries Education and Research Foundation with support from a federal grant. With products such as packaged goods, the UPC code gives manufacture and stocking unit information. For random weight products, the UPC code gives the type of item and the price information.

Sales activity in a supermarket generates interest in both the product and the supermarket itself. Many stores are trying to make shopping fun and make their store preferred by customers. Cooking demos and offerings of precooked samples give consumers a chance to try foods to which they might otherwise not have exposed themselves. It can be a very effective means of dealing with unusual products and new species. Videotapes playing at the fish counter area can also help fearful customers see how to handle this product properly. The recent development called the Cuisine Machine offers customers immediate, interactive computer access to information, including recipes, that can be printed out on-site. Many consumers do not know what to do with fish and appreciate assistance with suggestions on preparing the products. Recipes are often requested by consumers.

CONSUMER PERCEPTIONS AND BEHAVIOR

Understanding the customer is the essence of any successful marketing program. Several studies have uncovered some interesting facts with respect to consumer perceptions of fish. Two of these are discussed next.

Cornell University Survey (Bisogni et al. 1987)

- Customers do not trust their own purchase ability (knowledge) and thus depend on the person behind the counter. This encourages them to patronize small stores because consumers are not used to placing that kind of trust in supermarket personnel. Anything the supermarket can do to improve the "professional" standing of the fish counter staff would be a positive marketing tool. For example, they could distribute business cards with a "hot" phone line for help.
- Customers' perception of freshness is distorted. Consumers consider the first point of freshness to be the time the fish arrives at the store. A few days in the store and the fish is considered old. This perception can be a significant problem if the industry goes to shelf-life extension techniques.
- Customers do not like to handle the product or smell its cooked odor. Worse, customers do not realize the abuse that can occur in a car when the product is left out for an hour or so. They also do not know how to properly thaw or store the product in the home.
- Many customers overcook fish. Even those who understand the "10 minute" rule, that is, that cooking time should be based on thickness of product, worry about that fact that fish "tapers."

National Fisheries Institute/Better Homes and Gardens Biennial Surveys

These surveys take a close look every 2 years at the attitude of *Better Homes and Garden* readers to seafood. Although the selection of the sampled population may not accurately reflect the United States population, the surveys give a good general indicator of consumers' attitudes and are probably more accurate when it comes to detecting changes in these attitudes, since the same or similar questions are asked each time. For example, the main reasons for choosing fish are:

56% flavor/taste
55% health/nutrition
46% variety
73% who served fish considered it easy to prepare at home.

Notice though, that these results are drawn from a population that does cook fish.

- Six out of ten consumers have confidence in buying seafood, but almost 60% still prefer a full-service counter where they can talk to the fishmonger.
- Thirty-three percent of the population uses fish at least once a week; this is considered high compared to other surveys. A few years earlier, only 25% of the same panel members used fish once a week or more.
- About 35% thought that fresh and frozen fish were better than they were 5 years ago. 47% felt there was no change. Fewer than 8% believed they had become worse.
- Where do consumers eat fish? "Consumers who like fish are twice as likely to consume fish in restaurants as to cook it at home. However, even these consumers are unlikely to try a new species while eating out. Not very promising for new species! Those consumers who try new species generally like them, but even the daring want some menu suggestions. Another study suggested that the more people earn, the likelier they are to eat seafood away from home; and shellfish is the choice about half of the time.
- Consumer preference for form of cooking. (Note: more than one response is permitted per respondent.)

Broiled   88%
Baked   78%
Poached   21%
Frying   48% (and 34% opposed)

PRACTICAL MARKETING CONSIDERATIONS

The use of generic advertising can be extremely effective. It has been used to real advantage by the National Dairy Council and the National Livestock and Meat Board. Federal legislation was passed in 1988 that allowed for the formation of seafood marketing councils. The national council is responsible for providing broad generic programs and assisting in the development of both specific area or specific product councils. Funding was not immediately provided, but the bill should eventually provide funding for up to 4 years. After complaints from the Surgeon General of the United States, the industry's cartoon spokefish the "Sturgeon General" became the "fish spokesperson." The council has attempted to create a general awareness of fish, recommending fish meals twice a week. The council is due for renewal at the time of writing and it is not clear what the outcome will be. New individual product or regional councils have not emerged. Many in the industry seem to be questioning the effectiveness of this approach.

The marketplace is an extremely competitive environment, and expanding fish sales is a formidable undertaking. At present, the fresh fish counter is probably one of the least space-competitive spots in the supermarket; on the other hand, the frozen case is probably the most competitive. Each store has only so much frozen cabinet space: every new frozen product displaces an existing product. Few frozen products currently enjoy more than one facing.

To be successful takes serious planning. A second book by Ries and Trout (1986a) provides advice to companies with different positions in the marketplace. The important point is that each company must consider its marketplace competition in addition to considering its consumers. Basically the authors identify four positions:

1. The leader, should practice *defensive* marketing warfare along the following lines:

- Only the market leader should consider playing defense. One needs to evaluate one's position objectively!
- The best defensive strategy is the courage to go onto the attack. That is, to improve one's products constantly and to attempt to make it appear that the only competition is oneself.
- Strong competitive moves should always be blocked. That is, one should make one's weight felt and should not let the other companies take away market share. At the same time one should not waste one's effort trying to defend every last bit of territory thus risking spreading oneself too thin.

2. The other strong participants in the market should consider *offensive* marketing warfare:

- The main consideration is the strength of the leader's position. Understand who they are.
- Find a weakness in the leader's strength and attack at that point. Do not try to go head-to-head with the leader on the basis of their strength, exploit their weaknesses. Look at their strengths, which might preclude them from competing in the identified weakness. For example, the market leader might have the longest lines—the response is get people processed faster.
- Launch the attack on as narrow a front as possible. Put all of the effort into a battle that can be won because of strength in that narrowly focused area.

3.   A company with some strength should consider *flanking* marketing warfare:

- A good flanking move must be made into an uncontested area. The aim is to find an open niche that can be exploited.
- Tactical surprise ought to be an important element of the plan. It is important not to reveal your plans so that the competitors can move into the open niche.
- The pursuit is as critical as the attack itself. One needs the nerve to persevere and must provide adequate reserves. Cut losses and go with the winners.

4.   Finally, a small company might consider *guerrilla* marketing warfare:

- Find a segment of the market small enough to defend. Aim for the possible.
- No matter how much success is experienced, never act like the leader. Keep the organization lean and responsive.
- Be prepared to withdraw at a moment's notice. It is necessary to be flexible to succeed.

Despite the importance of consumer orientation in the actual products/ processes offered, this discussion emphasizes that the real battle is with competitors. A company must place itself in the right place with respect to these competitors. The more that is known about them and their plans and likely reactions, the better can moves be planned. Resources should be marshalled to fight one battle at a time, not to waste effort by being spread too thinly or by attacking head on.

A key pitfall is to expect to have better people—the bigger the organization the more statistically average they become; or to have a better product, the latter has to be perceived by the marketplace as such and that takes advertising and marketing.

Another problem is separation of strategy from tactics. The strategy must help to accomplish tactical needs. A good strategy will also not require brilliant tactics—if it does, it will not succeed.

Finally, a little luck never hurt anyone. Marketing fish may seem daunting, but the seafood industry needs newcomers with energy and innovation. In the words of Bel Kaufman (1964) in *Up the Down Staircase*: "Let it be a challenge to you."

REFERENCES

Aitken, A., I. M. Mackie, J. H. Merritt, and M. L. Windsor. 1982. *Fish Handling and Processing*. Edinburgh, Scotland: Her Majesty's Stationery Office.

Bisogni, C., G. A. Ryan, and J. M. Regenstein, 1987. What is fish quality? Can we incorporate consumer perceptions. In *Seafood Quality Determination*, D. E. Kramer and J. Liston, eds., pp. 547–563. Amsterdam: Elsevier.

Kaufman, B. 1964. *Up the Down Staircase*. Englewood Cliffs: Prentice Hall.

Ries, A., and J. Trout. 1986a. *Marketing Warfare*. New York: McGraw Hill.

Ries, A., and J. Trout. 1986b. *Positioning: The Battle for Your Mind*. New York: McGraw Hill.

16

Legal and Policy Issues

There are a number of government agencies that monitor the fish industry in the United States. At one time fish was the legal responsibility of the Bureau of Commercial Fisheries, a branch of the federal Department of the Interior. This placement was a recognition that issues concerning commercial fish were different from those related to sports fishing, but that the knowledge base for political and scientific/technical decisions often came from the agency's sports fishing base.

The Bureau was then transferred to the Department of Commerce and became the National Marine Fisheries Service (NMFS). It is still a part of the National Oceanic and Atmospheric Administration (NOAA), a department with a wide range of responsibilities including, for example, responsibility for weather satellites. Many members of the fishing and allied industries are opposed to this placement; they believe that fish and commercial fishing concerns are lost in an agency whose primary focus is international trade. It has often been suggested that fisheries belong under the jurisdiction of the United States Department of Agriculture (USDA), the nation's lead agency for aquaculture development.

Up to now, the NMFS has had FDA approval to administer specific programs of grading, performing inspections, and setting standards for fishery products. Related regulations are found in Title 50 of the *Code of Federal Regulations*. Current consideration of mandatory inspection (see below) might suggest that these activities belong under the jurisdiction of the USDA, rather than NMFS. The USDA has recently included fish more enthusiastically in its surplus foods purchasing and export programs. On the other hand, being a part of a separate agency has allowed fisheries to maintain an administrative structure that is free from competition with all the other major agricultural commodities.

As part of its effort to stimulate the fishing industry, the United States government has encouraged the formation of Fishery Development Foundations. These compete with other agencies and individuals for government funds from duties collected on imported fishery products and from the foreign

234

fishing fees. These "Saltonstall–Kennedy funds (SK funds)" provide money for fisheries development under the guidance of the industry. Participation in the development foundations is limited by membership dues. The amount of money made available each year for competitive grants is steadily decreasing. The funds go directly to NMFS to support on-going programs.

One of the legitimate roles of government is to set standards that buyers and sellers agree upon and which therefore aid the free flow of trade. These standards, that is, grades, and the definitions for these grades, must be drawn up in cooperation with the industry; however, the government must be meticulous in enforcing them fairly so that all parties can trust the grade assigned to any product. Grading serves various legitimate business purposes, but must be distinguished from inspection, which relates to health and safety concerns *per se*.

These grades give the consumer a more secure feeling about the quality of the product, for example, the Grade A program for fish. The key is to insure that the product is still Grade A at the time of purchase—not always easy with a perishable product like fish. Most standards are enforced earlier in the trade cycle, that is, prior to retail sale. Making inspections throughout the retail sector is an ambitious, and necessarily rare, undertaking. Current NMFS voluntary inspection programs do not include foreign fishery products. Some people feel that this program is designed to give American producers a market advantage; this policy may also discourage foreign producers from improving the quality of imported fish. Thankfully, in many cases the standards for fishery products are higher in other countries!

A "defect table" is used during the grading of fish products for which a standard exists. The grader subtracts points for various problems identified with the quality of the commodity. Many of these defects relate to workmanship, and do not affect the consumer; for example, uniformity of block-pack only concerns the manufacturer of further processed products.

The general comment "flavor and odor characteristic of the species" is usually the sole organoleptic basis for evaluating a product. This characteristic flavor and odor, whatever it may be, may not exist by the time the consumer purchases or consumes the product. Unfortunately, most grading and inspection occur too early in the market chain to effectively serve consumer needs.

Similar codes exist internationally either as official standards or as voluntary guidelines for good manufacturing practices. Many have been established by the Food and Agricultural Organization (FAO) and the World Health Organization (WHO) of the United Nations. The FAO is also responsible for the development of the Codex Alimentarius, essentially an international version of a trade standard with a large "Good Manufacturing Practice" component. For products not covered by any specific regulations, the United States FDA has established some general, but enforceable, rules on how to properly run a food business, the so-called official "Good Manufacturing Practices" that are included in Title 21 of the *Code of Federal Regulations*. A product can be seized or detained for failure to follow these guidelines. The FAO codes are based on the same operating concept and also deal with many of the same concerns as the FDA's. However, the FDA is not likely to adopt a Codex Alimentarius standard until American companies indicate more interest in participating or using such a method.

CODES AND LAWS

Every aspect of fish production can be monitored by one agency or another. Any local Board of Health may inspect the premises and workers of any food production or retail operation. The Departments of Labor, both the states' and the federal, monitor the economic status of workers as well as their health and safety; OSHA, the Occupational Safety and Health Administration, has become particularly active in this area. The Environmental Protection Agency and its equivalent state agencies monitor factory pollution. The revenue agencies, again both the states' and the federal, insure that taxes are paid. The general production of food, as mentioned previously, falls under the auspices of the Food and Drug Administration, a part of the Department of Health and Human Services.

The legal authority of all of these agencies is defined in legislation passed by both houses of Congress and usually signed by the President. The President, or the bill itself, then designates the agency that will administer the law. The appointed agency drafts a series of administrative codes, which are published in the *Federal Register*, a daily report of the actions of the executive or administrative branch of the United States government. The various agencies usually designate a period of time, usually one to six months, in which the public can comment on proposed changes in these administrative codes. Each agency generally records in the *Federal Register* what it has learned from the public and how it has taken the public's comments into account; its final ruling is recorded in the *Federal Register* as soon thereafter as possible. A separate set of books, the *Code of Federal Regulations* (CFR), is the repository for such changes of code. It contains all the official rules and regulations of the executive branch of the United States government; it is revised and reprinted every year. It consists of a total of 50 volumes, many of which run to over 1,000 pages. The four volumes dealing with food: USDA's Titles 7 and 9, FDA's Title 21, and NMFS's Title 50, take up about three feet of shelf space.

When, as a result of public comment, an agency believes that administrative codes must be changed significantly, the entire procedure may be completely started anew. Even after rules are included in the *Code of Federal Regulations*, they can be changed legally through the federal judicial system or by the agency. There is also a procedure for submitting a Citizen's Petition that allows any individual to attempt to change the law.

The material presented in the *Federal Register*—although difficult to read—can be extremely important to both businesses and consumers. It is often surprising how few comments are received on major regulatory matters. In a nation of 250 million people, although some issues generate thousands of letters, others may generate no letters at all.

Every issue of *Food Chemical News*, a weekly magazine (30–50 pages), highlights news in the areas of food chemicals, including those related to animal feeding. It follows the progress of various proposals through the legal system and indicates who is saying what about them.

The rules for the Food and Drug Administration are found in Title 21 of the *Code of Federal Regulations* (CFR), and the first three volumes (sections 1 through 199) include information on foods and food additives.

In addition to the FDA, the United States Department of Agriculture is actively involved in general food programs, including the inspection of meat and poultry, the grading of many commodities, the running of programs for school lunches and food stamps, and so on. Their rules are found in Titles 7 and 9 of the CFR. NMFS regulations are found in Title 50.

FISHERIES MANAGEMENT

In 1976 the United States declared a 200-mile exclusive fishing zone. This meant that, except for some highly migratory species (see later), the United States controls the fishing resources for 200 miles from every point of its coast. Before then, the United States had controlled only 12 miles. The first 3 miles were—and still are—under the jurisdiction of the individual states. The newly declared zone had previously been fished by boats from many different nations. At about the same time, Canada also declared a 200-mile limit. The two governments interpreted their new territories differently, and boundary disputes ensued. Unable to resolve the issues themselves, the two countries took the problem to the World Court, with both sides necessarily agreeing ahead of time to abide by its decision. In the interim, the two sides had also agreed to let fishermen from both countries fish in the disputed water.

In 1984 the court divided the disputed territory between the two countries. The United States lost some fishing grounds that were important to certain segments of the New England fleet, especially the scallop fleet. Curiously, the decision has displeased both countries. Unfortunately, the entire process has made it more difficult for the two countries to work together on other issues of common interest with respect to fisheries.

When the United States government first declared the 200-mile limit, it also tried to place the responsibility and authority for managing these resources into the hands of those most affected by the decision. Thus, several regional fisheries management councils were created. On the east coast, management is divided among three councils: the New England, the Mid-Atlantic, and the Southern and Gulf Councils. The North Pacific Council is responsible for the entire west coast, including Alaska and the western continental United States. There are serious problems to be dealt with in a timely manner. The councils must enforce controls on the industry in response to serious overfishing. This is extremely difficult to do because these limits immediately affect the livelihood of the fishing industry, often comprising neighbors and friends of the people trying to make these decisions.   Some of the tools available to the councils to control a catch are mesh size, other gear restrictions, official open and closed seasons, catch quotas by individual boat or trip, and quotas for the entire fishery. Councils can close off certain areas to fishing. In the future, they may also license fishermen or boats to catch certain amounts of fish. These limited-entry schemes would violate the common property aspect of fisheries that has been maintained in almost all American fisheries up to now. The licensing scheme would, however, give a fisherman the opportunity to spread the catch over a longer time period, for example, because if the fisherman is guaranteed the right to catch up to, say, 52,000 lb/year, it might be better to catch 1,000 lb per week all year, handle the fish well, and get a premium price. Another fisherman might use the entire allotment in Christmas week when many other fishermen want to be in port. It would permit fishermen to choose when to fish rather than for the "system" to make that decision.

The councils deal with many delicate issues relating to foreign fisheries. Interestingly, the pattern has been that the quality of international fishing relationships is inversely proportional to the quality of the United States government's overall relationships. In other words, the United States is involved in numerous fisheries controversies with its international "friends," but fisheries interactions have been smoothest where the two governments maintain serious ideological differences.

In dealing with foreign fisheries, the councils can put observers on foreign boats to insure that the quantity of fish, the species, and the by-catch are in accordance with United States law. Despite this observer program, controversies do arise. Many American fishermen believe that issues are brought to the State Department to be solved in terms of a greater political breadth that does not necessarily serve the American fishing industry well.

The United States has developed a policy nicknamed "Fish and Chips" that requires selection of a foreign partner based on its willingness (1) to share seafood technology with the United States, (2) to buy both at-sea and shoreside product, and (3) to cooperate with the development of the United

States fisheries. The goal is to eventually phase out the direct foreign allocations. In many fisheries, the United States has reached that point, so the controversial issues now deal with allocations for United States shoreside processing facilities versus United States processing at sea. The shoreside processors claim that they provide jobs and other local economic returns; the latter claim that they can operate more cheaply and provide a higher-quality fish by processing at sea.

In the meantime, the United States government does charge foreign boats for the "worth" of the fish caught at sea. Representative rates for at-sea harvesting of fish by foreign boats in US waters are given by NMFS as:

1984 duty rates per metric ton		*1980 duty rates per metric ton*	
Butterfish	$158	Atlantic shark	$164
Red hake	$29	Illex	$29
Silver hake	$31	Loligo	$110
River herring	$28		
Atlantic mackerel	$51		
Other finfish	$148		

Note that these species are underutilized in the United States, which explains why they are available as a direct catch to the foreign vessels.

The joint venture aspect of fisheries has tended to divide the industry. The foreign boats often provide a fixed price that holds even during gluts. On the other hand, shoreside processors feel that they cannot compete effectively for product with fixed prices.

The authority of the councils seems dubious. In the midst of a dynamic and intense environment, the councils' work is often slow and bureaucratic. In practice, one council is given the lead responsibility for a particular fishery or group of closely related fisheries. When it suggests proposals, the other councils within the same fishery are permitted to respond. The whole process is relatively slow; unfortunately, neither the fishermen nor the fish always want to wait for these wheels of government to churn. After prolonged work, the management councils send their policy suggestions to Washington; it is not uncommon for much time to pass before a final policy—often significantly modified—emerges. Each council is reauthorized annually, and the entire legal structure is revamped at intervals. In principle, these opportunities can be used to empower the councils. One serious fallacy in the system may be that there are too many international issues being dealt with by a "local council" that can never have the political leverage it needs.

Apparently, there are more interactions between fishing stocks than the current management system can deal with effectively on a species-by-species basis. It has been helpful to work with a group of fisheries at one time, but the process becomes noticeably more complex.

The information upon which business decisions of the councils are made sometimes requires technical data from various fishery scientists, many of whom are on the government payroll. It is difficult to verify reported data. For example, the biologists and the fishermen report different stock numbers: the research vessels can only make exploratory trips at certain times of year to certain grounds; the fishermen are out there all of the time!

Biologists are concerned with the recruitment of younger fish into the fishery. This is often discussed in terms of age classes, since each fish species tends to spawn once a year. The age and size of the fish at sexual maturity must be considered with respect to the behavior of the fishing gear, that is, smaller fish at sexual maturity are less likely to be caught. The councils are to use all these data to determine the maximum sustainable yield and allocate this available stock to the various competing American groups. What remains may then be offered for joint ventures with foreign processing vessels. Thereafter, foreign ships can be offered direct allotments. The total allowable foreign fishery (TALFF) is divided among competing countries. (Foreign fishing vessels with a direct allotment must pay for the actual amount of fish caught according to a schedule of charges established each year by NMFS.)

The reader may find a list of relevant abbreviations and terms to be helpful:

ABC (allowable biological catch): how many fish may be harvested in the given year.

IOY (initial optimum yield): the ideal number of fish to be harvested in the given year.

DAH (domestic annual harvest): how much the domestic fleet can handle of the given year's harvest.

DAP (domestic annual processing): how much of the given year's harvest the domestic processing industry can handle. Whether they expect to or not is not considered.

IJVP (initial joint venture processing): how much fish is available for joint ventures.

Reserve: fish held back for distribution later in the season.

TALFF (total allowable level of foreign fishing): what foreign vessels can fish.

FISHERIES MANAGEMENT IN CANADA

In Canada the federal government is responsible for making and executing all legal and management decisions concerning fisheries. The Canadian government is also committed to incorporating social policy into its plans, for example, preserving the fishing communities of Atlantic Canada. These commitments can conflict. The situation becomes more complicated with the distribution of both federal and provincial funding of Canadian fisheries. United States fishermen have charged that Canadian fisheries are subsidized to the point where Americans cannot compete in the United States market. (See later for a further discussion of Canadian subsidies.) This belief has led a group of New England fishing interests to apply for a countervailing duty policy through the International Trade Commission (ITC). The policy would make Canadian fishing companies pay a duty to bring their product into the United States; this duty would be equal to the "unfair" financial advantage the Canadian company has received from its government.

An important consideration arises: would this duty, if applied, be put on both fresh whole (H&G) fish and fillets or just on H&G fish? Most of the Canadian fresh fish imported into the United States are H&G; the fish are filleted and further processed in the United States. If the duty were placed only on H&G fish and not on fillets, it would serve to encourage Canadian fisheries to process more fish at home; further processing plants in New England would lose business.

The United States ITC has ruled that Canadian fish are unduly subsidized by their government and has imposed a temporary 6.8% duty on both fish and fillets. While this ruling was being contested, the payments on fish going from Canada to the United States were posted as a bond equal to the 6.8% duty; the monies became payable when the duty was finalized, a step that occurred at about the time this book was being written. An alternative for the Canadian government would be to collect a 6.8% export duty that would keep the funds in Canada.

The United States duty structure on frozen fish is such that raw fish blocks are generally imported and any breaded/battered products are manufactured in the United States. It is not surprising, then, that the two largest Canadian firms, Fishery Products and National Sea, both have processing plants in the United States. At this writing, the Canadian government owns about half of each of these processors and has a say in the management of the companies. After a few good years, following the government restructuring, it seems that both of these companies are again in difficulty.

Canadian Subsidies

An issue of great importance to the United States east coast industry is the question of the Canadian government's subsidies to its own fishing industry. Up until now, much of the Canadian fish harvested in Atlantic Canada has ended up as frozen product that is not in direct competition with higher-priced fresh fish.

In recent years, improvements in transportation have made it possible to bring fresh Canadian groundfish fillets into the United States; not surprisingly, the amount of Canadian fish coming into the United States has increased in recent years. Because of various economic programs provided by the Canadian government, these fish can be brought to Boston at lower prices than American fishermen feel they can afford to match. A restructuring of the fishing industry in Atlantic Canada has made the Canadian government a "virtual partner" in the business. Canadian efforts to upgrade quality standards and increase generic advertising are perceived by American processors and fishermen as an attempt to take over the United States fresh fish market. Various legal remedies are being explored and hearings are constantly in progress.

These controversies are not trivial. Fishing is often the only possible livelihood for many Canadians in the Maritimes and Newfoundland. As a result, fishing policy and the fishing subsidies are an integral and necessary part of life. In no state of the United States does fishing have as important a role in the life and economics of the people.

WITHDRAWAL PRICES

Special mention should be made of withdrawal prices. The European auction system has a European Economic Community (EEC)-established bottom price for any particular species of seafood. Any fish that is not sold above this withdrawal price will be purchased by the EEC and used for noncompetitive use, for example, for donations to hospitals, prisons, and so on. This encourages fishermen to participate in the auction process without fear that they will not be able to sell their fish.

SPORTS FISHING

Sports fishermen and commercial fishermen are in constant competition for fish. They are also in competition for a political voice. Since one commercial fisherman has the same voting strength as one sports fisherman, the sports fishermen are often a stronger political entity at the state level. Thus, although the commercial fishermen serve the entire population, the laws often tend to favor the sports fishermen.

Sports fishermen claim that the commercial fishermen take too many fish. Commercial fishermen—who rely on earnings as their livelihood—are incensed that sports fishermen might sell their catch to restaurants that used to buy from commercial fishermen. States such as New York have passed laws to limit this practice.

In many states, particularly in the Midwest, commercial fishing has almost become extinct. This is less surprising when we read negative rhetoric like this item from the Bulletin of the Illinois Salmon Unlimited Legislative Report:

> There is no room in government today for legislators or bureaucrats that are unresponsive to the needs of the sportsmen and outdoor recreationists. We believe the voting sportsmen should and will assure the replacement of those officials that do not support our needs! . . . . We are opposed to a commercial fishery in the Illinois portion of Lake Michigan. Salmon Unlimited and the Illinois Sportsmen's Legislative Coalition are unalterably opposed to the continuation of a commercial fishery in the Illinois portion of Lake Michigan. We support the elimination of the five existing licenses/permits for commercial fishing through attrition. . . .

Sports and commercial fishermen have much to gain by working together. There are environmental issues that are critical to the success of both groups—and to the health and well-being of all the people and the earth. One can only hope that the future will see more efforts at cooperation than conflict between these two groups.

KOSHER VERSUS NONKOSHER FISH

The distinction between acceptable and unacceptable fish with respect to the Jewish dietary laws is that kosher fish must have fins and scales. "Scales" are defined as a covering that can be removed from the fish without tearing the skin, generally referring to ctenoid or cycloid scales. Fish with ganoid or placoid scales—even within the pisces group of fish—are generally not considered kosher; for example, shark, monkfish, lumpfish, catfish, sturgeon, and for the traditional Jews, the swordfish. All other marine animals are not kosher, including all shellfish.

Note that as part of the turbot family, nonkosher European turbot (*Scophthalmus maximus* or *Psetta maximus*) may be sold with other turbots that are kosher. Another fish in this category is ocean pout, a nonkosher species that is sometimes sold at retail with kosher flatfish.

Within the kosher laws, fish is not considered a meat. Therefore, the laws requiring the separation of milk and milk products from meat do not apply to fish. A cheeseburger or a mined taco filling with grated cheese on top are prohibited when made with beef or poultry; however, they would be perfectly kosher if made with (kosher) fish. This fact suggests some interesting product development projects within the expanding kosher food market. In the past decade, it has been estimated that the number of kosher products has increased from 1,000 to 18,000 and that as many as 30% of the products in a supermarket in the Northeastern United States may be kosher-approved. The market for these products goes beyond the Jewish kosher consumer, including vegetarians, Moslems, Seventh Day Adventists, and many other consumers who use the kosher certifications as a symbol of quality. The following list (prepared by James W. Atz and reproduced by courtesy of the Union of Orthodox Jewish Congregations of America) gives the kosher and nonkosher species of fish found in the United States. For more details on this subject, consult Regenstein and Regenstein (1979, 1988).

The two lists cover most of the fishes sold for food or angled for sport in the United States of America. It would be almost impossible to prepare a list that includes all such fishes, because new ones may appear on the market at any time.

In order to make these lists useful to the layman, they had to be based on the popular names of the fishes, scientific names being included only to help the expert check identifications. But the popular names of fishes have always been a source of error and confusion. The white perch is not a perch, for example, and very different species are sometimes given the same vernacular name. Nevertheless, with reasonable care, serious mistakes should not occur, and a nonkosher species should never be taken for a kosher one. The numerous cross-references will insure that none of the different kinds of fishes bearing a particular name will be missed, and in all but two instances (the jacks and the flounders) in which the Family is listed as kosher, it may be safely assumed that all its members are properly finned and scaled.

Kosher Fishes

Albacore. See: Mackerels

Alewife. See: Herrings

Amberjack. See: Jacks

Anchovies (Family Engraulidae)
  Including:
    European anchovy (*Engraulis encrasicolus*)
    Northern or California anchovy (*Engraulis mordax*)

Angelfishes and butterfly fishes (Family Chaetodontidae)
  Including:
    Angelfishes (*Holacanthus* species, *Pomacanthus* species)

Angler. See: Goosefishes (nonkosher)

Atlantic Pomfret or Ray's bream (*Brama brama*)

Ballyhoo. See: Flyingfishes

Barracudas (Family Sphyraenidae)
  Including:
    Barracudas and kakus (*Sphyraena* species)

Bass. See: Sea basses. Temperate basses. Sunfishes. Drums.

Beluga. See: Sturgeons (nonkosher)

Bigeyes (Family Priacanthidae)
  Including:
    Bigeyes or aweoweos (*Priacanthus* species)

Blackfish. See: Carps. Wrasses
Blacksmith. See: Damselfishes
Blueback. See: Flounders. Herrings. Trouts
Bluefish or snapper blue (*Pomatomus saltatrix*)
Bluegill. See: Sunfishes
Blowfish. See: Puffers (nonkosher)
Bocaccio: See: Scorpionfishes
Bombay duck (*Harpadon nehereus*)
Bonefish (*Albula vulpes*)
Bonito. See: Cobia. Mackerels
Bowfin, freshwater dogfish, or grindle (*Amia calva*)
Bream. See: Carps. Atlantic pomfret. Porgies
Brill. See: Flounders
Buffalo fishes. See: Suckers
Bullhead. See: Catfishes (nonkosher)
Burbot. See: Codfishes
Butterfishes (Family Stromateidae)
Including:
Butterfish (*Peprilus triacanthus*)
Pacific pompano (*Peprilus simillimus*)
Harvestfishes (*Peprilus* species)
Butterfly fish. See: Angelfish
Cabezon. See: Sculpins (nonkosher)
Cabrilla. See: Sea basses
Calico bass. See: Sunfishes
Capelin. See: Smelts
Carps and minnows (Family Cyprinidae)
Including:
The carp, leather carp, mirror carp (*Cyprinus carpio*)
Crucian carp (*Carassius carassius*)
Goldfish (*Carassius auratus*)
Tench (*Tinca tinca*)
Splittail (*Pogonichthys macrolepidotus*)

Squawfishes (*Ptychocheilus* species)
Sacramento blackfish or hardhead (*Orthodon microlepidotus*)
Freshwater breams (*Abramis* species, *Blicca* species)
Roach (*Rutilus rutilus*)
Carpsucker. See: Suckers
Caviar. See: Trouts and whitefishes (salmon). Lumpsuckers (nonkosher). Sturgeons (nonkosher)
Cero. See: Mackerels
Channel bass. See: Drums
Char. See: Trouts
Chilipepper. See: Scorpionfishes
Chinook salmon. See: Trouts
Chub. See: Trouts. Sea chubs
Cichlids (Family Cichlidae)
Including:
Tilapias (*Tilapia* species)
Mozambique mouthbrooder (*Tilapia mossambica*)
Cichlids (*Cichlasoma* species)
Rio Grande perch (*Cichlasoma cyanoguttatum*)
Cigarfish. See: Jacks
Cisco. See: Trouts
Coalfish. See: Codfishes
Cobia, cabio, or black bonito (*Rachycentron canadum*)
Cod, cultus, black, blue, or ling. See: Greenlings. Sablefish
Codfishes (Family Gadidae)
Including:
Cod (*Gadus morhua*)
Haddock (*Melanogrammus aeglefinus*)
Pacific cod (*Gadus macrocephalus*)
Pollock, saithe, or coalfish (*Pollachius virens*)

Walleye pollock (*Theragra chalcogramma*)
Hakes (*Urophycis* species)
Whiting (*Merlangius merlangus*)
Blue whiting or poutassou (*Micromesistius poutassou*)
Burbot, lawyer, or freshwater ling (*Lota lota*)
Tomcods or frostfishes (*Microgradus* species)
Coho salmon. See: Trouts
Corbina or corvina. See: Drums
Cottonwick. See: Grunts
Crappie. See: Sunfishes
Crevalle. See: Jacks
Croaker. See: Drums
Crucian carp. See: Carps
Cubbyu. See: Drums
Cunner. See: Wrasses
Dab. See: Flounders
Damselfishes (Family Pomacentridae)
Including:
Blacksmith (*Chromis punctipinnis*)
Garibaldi (*Hypsypops rubicunda*)
Doctorfish. See: Surgeonfishes
Dogfish. See: Bowfin. Sharks (nonkosher)
Dolly Varden. See: Trouts
*Dolphin fishes or mahimahi (*Coryphaena* species)
Drums and croakers (Family Sciaenidae)
Including:
Seatrouts and corvinas (*Cyanoscion* species)
Weakfish (*Cynoscion nebulosus*)

White seabass (*Cynoscion nobilis*)
Croakers (*Micropogon* species, *Bairdiella* species, *Odontoscion* species)
Silver perch (*Bairdiella chrysura*)
White or king croaker (*Genyonemus lineatus*)
Black croaker (*Cheilotrema saturnum*)
Spotfin croaker (*Roncador stearnsi*)
Yellowfin croaker (*Umbrina roncador*)
Drums (*Pogonias* species, *Stellifer* species, *Umbrina* species)
Red drum or channel bass (*Sciaenops ocellata*)
Freshwater drum (*Aplodinotus grunniens*)
Kingfishes or king whitings (*Menticirrhus* species)
California corbina (*Menticirrhus undulatus*)
Spot or lafayette (*Leiostomus xanthurus*)
Queenfish (*Seriphus politus*)
Cubbyu or ribbon fish (*Equetus umbrosus*)
Eulachon. See: Smelts
Flounders (Families Bothidae and Pleuronectidae)
Including:
Flounders (*Paralichthys* species, *Liopsetta* species, *Platichthys* species, etc.)
Starry flounder (*Platichthys stellatus*)

---

* Not to be confused with the mammal called Dolphin or Porpoises which is nonkosher..

Summer flounder or fluke
(*Paralichthys dentatus*)
Yellowtail flounder (*Limanda ferrugina*)
Winter flounder, lemon sole, or blackback
(*Pseudopleuronectes americanus*)
Halibuts (*Hippoglossus* species)
California halibut
(*Paralichthys californicus*)
Bigmouth sole (*Hippoglossina stomata*)
Butter or scalyfin sole
(*Isopsetta isolepis*)
"Dover" sole (*Microstomus pacificus*)
"English" sole (*Parophrys vetulus*)
Fantail sole (*Xystreurys liolepis*)
Petrale sole (*Eopsetta jordani*)
Rex sole (*Glyptocephalus zachirus*)
Rock sole (*Lepidopsetta bilineata*)
Sand sole (*Psettichthys melanostictus*)
Slender sole (*Lyopsetta exilis*)
Yellowfin sole (*Limanda aspera*)
Pacific turbots (*Pleuronichthys* species)
Curlfin turbot or sole
(*Pleuronichthys decurrens*)
Diamond turbot (*Hypsopsetta guttulata*)
Greenland turbot or halibut
(*Reinhardtius hippoglossoides*)
Sanddabs (*Citharichthys* species)
Dabs (*Limanda* species)
American plaice
(*Hippoglossoides platessoides*)

European plaice (*Pleuronectes platessa*)
Brill (*Scophthalmus rhombus*)
But not including:
European turbot (*Scophthalmus maximus* or *Psetta maximus*)
Fluke. See: Flounders
Flyingfishes and halfbeaks (Family Exocoetidae)
Flyingfishes (*Cypselurus* species, and others)
Ballyhoo or balao (*Hemiramphus* species)
Frostfish. See: Codfishes
Gag. See: Sea basses
Gar. See: Needlefishes. Gars (nonkosher)
Garibaldi. See: Damselfishes
Giant kelpfish (*Heterostichus rostratus*)
Gizzard shad. See: Herrings
Goatfishes or surmullets (Family Mullidae)
Including:
Goatfishes (*Mullus* species, *Pseudopeneus* species); Wekes or goatfishes (*Mulloidichthys* species, *Upeneus* species)
Kumu (*Parupeneus* species)
Red mullet (*Mullus surmuletus*)
Gobies (Family Gobiidae)
Including:
Bigmouth sleeper or guavina (*Gobiomorus dormitor*)
Sirajo goby (*Sicydium plumieri*)
Goldeye and mooneye (*Hiodon alosoides* and *Hiodon tergisus*)
Goldfish. See: Carps
Grayfish. See: Sharks (nonkosher)
Grayling. See: Trouts
Graysby. See: Sea basses
Greenlings (Family Hexagrammidae)

Including:
  Greenlings (*Hexagrammos*
    species)
  Kelp greenling or seatrout
    (*Hexagrammos decagrammus*)
  Lingcod, cultus or blue cod
    (*Ophiodon elongatus*)
  Atka mackerel (*Pleurogrammus
    monopterygius*)
Grindle. See: Bowfin
Grouper. See: Sea basses
Grunion. See: Silversides
Grunts (Family Pomadasyidae)
  Including:
    Grunts (*Haemulon* species,
      Pomadasys species)
    Margate (*Haemulon album*)
    Tomtate (*Haemulon
      aurolineatum*)
    Cottonwick (*Haemulon
      melanurum*)
    Sailors choice (*Haemulon
      parrai*)
    Porkfish (*Anisotremus
      virginicus*)
    Black margate (*Anisotremus
      surinamensis*)
    Sargo (*Anisotremus davidsoni*)
    Pigfish (*Orthopristis
      chrysoptera*)
Guavina. See: Gobies
Haddock. See: Codfishes
Hake. See also: Codfishes
Hakes (Family Merlucciidae)
  Including:
    Hakes (*Merluccius* species)
    Silver hake or whiting
      (*Merluccius bilinearis*)
    Pacific hake or merluccio
      (*Merluccius productus*)
Halfbeak. See: Flyingfishes
Halfmoon. See: Sea chubs
Halibut. See: Flounders

Hamlet. See: Sea basses
Hardhead. See: Carps
Harvestfish. See: Butterfishes
Hawkfishes (Family Cirrhitidae)
  Including:
    Hawkfishes (*Cirrhitus* species)
Herrings (Family Clupeidae)
  Including:
    Atlantic and Pacific herring
      (*Clupea harengus* subspecies)
    Thread herrings (*Opisthonema*
      species)
    Shads (*Alosa* species)
    Shad or glut herring, or
      blueback (*Alosa aestivalis*)
    Hickory shad (*Alosa mediocris*)
    Alewife or river herring
      (*Alosa pseudoharengus*)
    Gizzard shads (*Dorosoma*
      species)
    Menhadens or mossbunkers
      (*Brevoortia* species)
    Spanish sardine (*Sardinella
      anchovia*)
    European sardine or pilchard
      (*Sardina pilchardus*)
    Pacific sardine or pilchard
      (*Sardinops sagax*)
    Sprat (*Sprattus sprattus*)
Hind. See: Sea basses
Hogchoker. See: Soles
Hogfish. See: Wrasses
Horse mackerel. See: Jacks
Jack mackerel. See: Jacks
Jacks and pompanos (Family
  Carangidae)
  Including:
    Pompanos, palometas, and
      permits (*Trachinotus* species)
    Amberjacks and yellowtails
      (*Seriola* species)
    California yellowtail (*Seriola
      dorsalis*)

Scads and cigarfish (*Decapterus*
species, *Selar* species,
Trachurus species)
Jack mackerel or horse
mackerel (*Trachurus
symmetricus*)
Jacks and uluas (*Caranx*
species, *Carangoides* species)
Crevalles (*Caranx* species)
Blue runner (*Caranx crysos*)
Rainbow runner (*Elagatis
bipinnulata*)
Moonfishes (*Vomer* species)
Lookdown (*Selene vomer*)
Leatherback or lae
(*Scomberoides sanctipetri*)
But not including:
Leatherjacket (*Oligoplites
saurus*)
Jacksmelt. See: Silversides
Jewfish. See: Sea basses
John Dory (*Zeus faber*)
Kelpfish. See: Giant kelpfish
Kingfish. See: Drums. Mackerels
Ladyfish, or tenpounder (*Elops
saurus*)
Lafayette. See: Drums
Lake herring. See: Trouts
Lance or launce. See: Sand lances
Largemouth bass. See: Sunfishes
Lawyer. See: Codfishes
Leatherback. See: Jacks
Leatherjacket. See: Jacks
(nonkosher)
Lingcod. See: Greenlings
Lizardfishes (Family Synodontidae)
Lookdown. See: Jacks
Mackerel. See also: Jacks
Mackerel, Atka. See: Greenlings
Mackerels and tunas (Family
Scombridae)
Including:
Mackerels (*Scomber* species,

Scomberomorus species,
Auxis species)
Spanish mackerels, cero, and
sierra (*Scomberomorus*
species)
King mackerel or kingfish
(*Scomberomorus cavalla*)
Bonitos (*Sarda* species)
Wahoo (*Acanthocybium
solanderi*)
Tunas (*Thunnus* species,
Euthynnus species)
Skipjack tunas (*Euthynnus* or
Katsuwonus species)
Albacore (*Thunnus alalunga*)
But not including:
Snake mackerels
Mahimahi. See: Dolphin fishes
Margate. See: Grunts
Marlin. See: Billfishes (nonkosher)
Menhaden. See: Herrings
Menpachii. See: Squirrelfishes
Merluccio. See: Hakes
Midshipman. See: Toadfishes
(nonkosher)
Milkfish or awa (*Chanos chanos*)
Mojarras (Family Gerreidae)
Including:
Mojarras (*Eucinostomus* species,
Guerres species, *Diapterus*
species)
Monkeyface prickleback or eel
(*Cebidichthys violaceus*)
Mooneye. See: Goldeye
Moonfish. See: Jacks
Mossbunker. See: Herrings
Mouthbrooder. See: Cichlids
Mullet. See: Goatfishes
Mullets (Family Mugilidae)
Including:
Mullets and amaamas (*Mugil*
species)
Uouoa (*Neomyxus chaptalii*)

Mountain mullets or dajaos
(*Agonostomus* species)
Muskellunge. See: Pikes
Mutton hamlet. See: Sea basses
Muttonfish. See: Snappers
Needlefishes (Family Belonidae)
Needlefishes or marine gars
(*Strongylura* species, *Tylosurus*
species)
Opaleye. See: Sea chubs
Paddlefish. See: Sturgeons
(nonkosher)
Palometa. See: Jacks
Parrotfishes (Family Scaridae)
Including:
Parrotfishes and uhus (*Scarus*
species, *Sparisoma* species)
Perch. See also: Temperate basses.
Drums. Cichlids. Surfperches.
Scorpionfishes
Perches (Family Percidae)
Including:
Yellow perch (*Perca flavescens*)
Walleye, pike-perch, or yellow
or blue pike (*Stizostedion
vitreum*)
Sauger (*Stizostedion canadense*)
Permit. See: Jacks
Pickerel. See: Pike
Pigfish. See: Grunts
Pike. See also: Perches
Pikes (Family Esocidae)
Including:
Pike (*Esox lucius*)
Pickerels (*Esox* species)
Muskellunge (*Esox
masquinongy*)
Pike-perch. See: Perches
Pilchard. See: Herrings
Pinfish. See: Porgies
Plaice. See: Flounders
Pollock. See: Codfishes
Pomfret. See: Atlantic pomfret

Pompano. See: Jacks. Butterfishes
Porgies and sea breams (Family
Sparidae)
Including:
Porgies (*Calamus* species,
Diplodus species, *Pagrus*
species)
Scup (*Stenotomus chrysops*)
Pinfish (*Lagodon rhomboides*)
Sheepshead (*Archosargus
probatocephalus*)
Porkfish. See: Grunts
Pout. See: Ocean pout (nonkosher)
Poutassou. See: Codfishes
Prickleback. See: Monkeyface
prickleback. Rockprickleback
(nonkosher)
Queenfish. See: Drums
Quillback. See: Suckers
Rabalo. See: Snooks
Ratfish. See: Sharks (nonkosher)
Ray. See: Sharks (nonkosher)
Ray's bream. See: Atlantic pomfret
Red snapper. See: Snappers
Redfish. See: Scorpionfishes. Wrasses
Roach. See: Carps
Rock bass. See: Sunfishes
Rock hind. See: Sea basses
Rockfish. See: Scorpionfishes.
Temperate basses
Rosefish. See: Scorpionfishes
Rudderfish. See: Sea chubs
Runner. See: Jacks
Sablefish or black cod (*Anoplopoma
fimbria*)
Sailfish. See: Billfishes (nonkosher)
Sailors choice. See: Grunts
Saithe. See: Codfishes
Salmon. See: Trouts
Sand lances, launces, or eels
(*Ammodytes* species)
Sardine. See: Herrings
Sargo. See: Grunts

Sauger. See: Perches

Scad. See: Jacks

Scamp. See: Sea basses

Schoolmaster. See: Snappers

Scorpionfishes (Family
   Scorpaenidae)
   Including:
      Scorpionfishes (*Scorpaena*
         species)
      California scorpionfish or
         sculpin (*Scorpaena guttata*)
      Nohus (*Scorpaenopsis* species)
      Redfish, rosefish, or ocean
         perch (*Sebastes marinus*)
      Rockfishes (*Sebastes* species,
         Sebastodes species)
      Pacific ocean perch (*Sebastes
         alutus*)
      Chilipepper (*Sebastes goodei*)
      Bocaccio (*Sebastes
         paucispinus*)
      Shortspine thornyhead or
         channel rockfish
         (*Sebastolobus alascanus*)

Scup. See: Porgies

Sea bass. See also: Temperate
   basses. Drums

Sea basses (Family Serranidae)
   Including:
      Black sea basses (*Centropristis*
         species)
      Groupers (*Epinephelus* species
         and *Mycteroperca* species)
      Rock hind (*Epinephelus
         adscensionis*)
      Speckled hind (*Epinephelus
         drummondhayi*)
      Red hind (*Epinephelus guttatus*)
      Jewfish (*Epinephelus itajara*)
      Spotted cabrilla (*Epinephelus
         analogus*)
      Gag (*Mycteroperca microlepis*)
      Scamp (*Mycteroperca phenax*)

Graysby (*Petrometopon
   cruentatum*)
Mutton hamlet (*Alphestes afer*)
Sand bass, kelp bass, and spotted
   bass (*Paralabrax* species)

Sea bream. See: Porgies

Sea chubs (Family Kyphosidae)
   Including:
      Bermuda chub or rudderfish
         (*Kyphosus sectatrix*)
      Opaleye (*Girella nigricans*)
      Halfmoon (*Medialuna
         californiensis*)

Seaperch. See: Surfperches

Searaven. See: Sculpins (nonkosher)

Searobins (Family Triglidae)
      Searobins (*Prionotus* species)

Sea-squab. See: Puffers (nonkosher)

Seatrout. See: Drums. Greenlings.
   Steelhead

Shad. See: Herrings

Sheepshead. See: Porgies. Wrasses

Sierra. See: Mackerels

Silversides (Family Atherinidae)
   Including:
      Whitebait, spearing, or
         silversides (*Menidia* species)
      California grunion (*Leuresthes
         tenuis*)
      Jacksmelt (*Atherinopsis
         californiensis*)
      Topsmelt (*Atherinops affinis*)

Sirajo goby. See: Gobies

Skates. See: Sharks (nonkosher)

Skipjack. See: Mackerels

Sleeper. See: Gobies

Smallmouth bass. See: Sunfishes

Smelts (Family Osmeridae)
   Including:
      Smelts (*Osmerus* species)
      Capelin (*Mallotus villosus*)
      Eulachon (*Thaleichthys
         pacificus*)

Shiner perch (*Cymatogaster aggregata*)
Surgeonfishes (Family Acanthuridae)
Including:
    Surgeonfishes and tangs (*Acanthurus* species, *Zebrasoma* species)
    Doctorfish (*Acanthurus chirurgus*)
    Unicornfishes or kalas (*Naso* species)
Tang. See: Surgeonfishes
Tarpon (*Megalops atlantica*)
Tautog. See: Wrasses
Temperate basses (Family Percichthyidae)
Including:
    Striped bass or rockfish (*Morone sazatilis*)
    Yellow bass (*Morone mississippiensis*)
    White bass (*Morone chrysops*)
    White perch (*Morone americana*)
    Giant California sea bass (*Stereolepis gigas*)
Tench. See: Carps
Tenpounder. See: Ladyfish
Threadfins (Family Polynemidae)
Including:
    Blue bobo (*Polydactylus approximans*)
    Barbu (*Polydactylus virginicus*); Moi (*Polydactylus sexfilis*)
Tilapia. See: Cichlids.
Tilefishes (Family Branchiostegidae)
Including:
    Tilefish (*Lopholatilus chamaeleonticeps*)
    Ocean whitefish (*Caulolatilus princeps*)
Tomcod. See: Codfishes

Tomtate. See: Grunts
Topsmelt. See: Silversides
Tripletail (*Lobotes surinamensis*)
Trouts and whitefishes (Family Salmonidae)
Including:
    Atlantic salmon (*Salmo salar*)
    Pacific salmons (*Oncorhynchus* species) Coho or silver salmon; sockeye, blueback or red salmon; chinook, king or spring salmon; pink or humpback salmon; chum, dog, or fall salmon
    Trouts (*Salmo* species) Brown trout, rainbow trout or steelhead, cutthroat trout, golden trout
    Chars (*Salvelinus* species) Lake trout, brook trout, Arctic char, Dolly Varden
    Whitefishes and ciscos (*Coregonus* species and *Prosopium* species)
    Cisco or lake herring (*Coregonus artedii*)
    Chubs (*Coregonus* species)
    Graylings (*Thymallus* species)
Tuna. See: Mackerels
Turbot. See: Flounders (some nonkosher)
Unicornfish. See: Surgeonfishes
Wahoo. See: Mackerels
Walleye. See: Perches
Walleye pollock. See: Codfishes
Warmouth. See: Sunfishes
Weakfish. See: Drums
Whitebait. See: Silversides
Whitefish. See: Trouts. Tilefishes
Whiting. See: Codfishes. Hakes. Drums
Wrasses (Family Labridae)
Including:

Hogfishes and aawas (*Bodianus* species)
Hogfish or capitaine (*Lachnolaimus maximus*)
Tautog or blackfish (*Tautoga onitis*)

California sheephead or redfish (*Pimelometopon pulchrum*)
Cunner, chogset, or bergall (*Tautogolabrus adspersus*)
Yellowtail. See: Jacks
Yellowtail snapper. See: Snappers

Nonkosher fishes

Billfishes (Family Istiophoridae)
Including:
Sailfishes (*Istiophorus* species)
Marlins and spearfishes (*Tetrapterus* species, *Makaira* species)
Catfishes (Order Siluriformes)
Including:
Channel catfish (*Ictalurus punctatus*)
Bullheads (*Ictalurus* species)
Sea catfish (*Arius felis*)
Cutlassfishes (Family Trichiuridae)
Including:
Cutlassfishes (*Trichiurus* species)
Scabbardfishes (*Lepidopus* species)
Eels (Order Anguilliformes)
Including:
American and European eel (*Anguilla rostrata* and *Anguilla anguilla*)
Conger eel (*Conger oceanicus*)
Gars (Order Semionotiformes)
Freshwater gars (*Lepisosteus* species)
Goosefishes or anglers (*Lophius* species)
Lampreys (Family Petromyzontidae)
Leatherjacket (*Oligoplites saurus*)

Lumpsuckers (Family Cyclopteridae)
Including:
Lumpfish (*Cyclopterus lumpus*)
Snailfishes (*Liparis* species)
Ocean pout or eelpout (*Macrozoarces americanus*)
Oilfish (*Ruvettus pretiosus*)
Puffers (Family Tetraodontidae)
Puffers, blowfishes, swellfishes, sea-squab (*Sphoeroides* species)
Rock prickleback or rockeel (*Xiphister mucosus*)
Sculpins (Family Cottidae)
Including:
Sculpins (*Myoxocephalus* species, *Cottus* species, *Leptocottus* species, etc.)
Cabezon (*Scorpaenichthys marmoratus*)
Searaven (*Hemitripterus americanus*)
Sharks, rays, and their relatives (Class Chondrichthyes)
Including:
Grayfishes or dogfishes (*Mustelus* species, *Squalus* species)
Soupfin shark (*Galeorhinus zyopterus*)
Sawfishes (*Pristis* species)
Skates (*Raja* species)

Chimaeras or ratfishes (Order
Chimaeriformes)
Snake mackerels (*Gempylus*
species)
Sturgeons (Order Acipenseriformes)
Including:
Sturgeons (*Acipenser* species,
Scaphirhynchus species)
Beluga (*Huso huso*)
Paddlefish or spoon bill cat
(*Polyodon spathula*)
Swordfish (*Xiphias gladius*)
Toadfishes (Family Batrachoididae)
Including:

Toadfishes (*Opsanus* species)
Midshipmen (*Porichthys*
species)
Triggerfishes and filefishes (Family
Balistidae)
Triggerfishes (*Balistes* species,
Canthidermis species)
Trunkfishes (Family Ostraciidae)
Trunkfishes and cowfishes
(*Lactophrys* species)
Wolffishes (Family Anarhichadidae)
Including:
Wolffishes or ocean catfishes
(*Anarhichas* species)

REFERENCES

Regenstein, J. M., and C. E. Regenstein. 1979. An introduction to the kosher (dietary) laws for food scientists and food processors. *Food Technol.* **33** (1): 89–99.

Regenstein, J. M., and C. E. Regenstein. 1988. The kosher dietary laws and their implementation in the food industry. *Food Technol.* **42**(6): 86–94.

Bibliography

Aitken, A., I. M. Mackie, J. H. Merritt, and M. L. Windsor. 1982. *Fish Handling and Processing*. Aberdeen: Ministry of Agriculture, Fisheries, and Foods, Torry Research Station, and Edinburgh: Her Majesty's Stationery Office.

Clucas, I. J. (Compiler). 1982. *Fish Handling, Preservation and Processing in the Tropics*. Part 2. London: Report of the Tropical Products Institute.

Chaston, I. 1982. *Marketing in Fisheries and Aquaculture*. Farnham, Surrey, England: Fishing News Books Ltd.

Connell, J. J. 1980. *Advances in Fish Science and Technology*. Farnham, Surrey, England: Fishing News Books Ltd.

Connell, J. J. 1980. *Control of Fish Quality*, 2d Edition. Farnham, Surrey, England: Fishing News Books Ltd.

Faria, S. 1984. *The Northeast Seafood Book, A Manual of Seafood Products, Marketing and Utilization*. Boston: Massachusetts Division of Marine Fisheries.

International Institute of Refrigeration. 1982. *Advances in Technology in the Chilling, Freezing, Processing, Storage, and Transport of Fish, especially Underutilized Species*. Paris: Commissions C2, D1, D2, and D3.

Jarvis, N. D. 1943. *Principles and Methods in the Canning of Fishery Products*. Washington: United States Department of the Interior. Research Report 7.

Jarvis, N. D. 1950. *Curing of Fishery Products*. Washington: United States Department of the Interior. Research Report 18.

Kramer, D. E., and J. Liston (eds.).1987. *Seafood Quality Determination*. Amsterdam: Elsevier.

Lands, W. E. M. 1986. *Fish and Human Health*. Orlando, FL: Academic Press.

Love, R. M. 1970. *The Chemical Biology of Fishes*. New York: Academic Press.

Love, R. M. 1980. *The Chemical Biology of Fishes*, Vol. II. New York: Academic Press.

Love, R. M. 1988. *The Food Fishes: Their Intrinsic Variation and Practical Implications*. New York: Van Nostrand Reinhold.

Martin, R. E. 1980. *Third National Technical Seminar on Mechanical Recovery and Utilization of Fish Flesh*. Washington: National Fisheries Institute.

Martin, R. E. 1982. *Proceedings of the First National Conference on Seafood Packaging and Shipping*. Washington: National Fisheries Institute.

Martin, R. E., and G. J. Flick. 1990. *The Seafood Industry*. New York: Van Nostrand Reinhold.

Martin, R. E., G. J. Flick, C. E. Hebard, and D. R. Ward. 1982. *Chemistry and Biochemistry of Marine Food Products.* Westport, CT: Avi.

Meade, J. W. 1989. *Aquaculture Management.* New York: Van Nostrand Reinhold.

National Research Council. 1985. *An Evaluation of the Role of Microbiological Criteria for Foods and Food Ingredients.* Washington: National Academy Press.

Nettleton, J. A. 1985. *Seafood Nutrition: Facts, Issues and Marketing of Nutrition in Fish and Shellfish.* Huntington, NY: Osprey Books.

Nettleton, J. A. 1987. *Seafood and Health.* Huntington, NY: Osprey Books.

Novikov, V. M. (ed.). 1981. *Handbook of Fishery Technology.* New Delhi: Amerind Publishing Co. Pvt. Ltd.

Piper, R. G., I. B. McElwain, L. E. Orme, J. P. McCraren, L. G. Fowler, and J. R. Leonard. 1982. *Fish Hatchery Management.* Washington: United States Department of the Interior, Fish and Wildlife Service.

Ries, A., and J. Trout. 1989. *Bottom-Up Marketing.* New York: McGraw Hill.

Stansby, M. 1990. *Fish Oils in Nutrition.* New York: Van Nostrand Reinhold.

Suzuki, T. 1981. *Fish and Krill Proteins, Processing Technology.* London: Applied Science Publishers.

University of North Carolina Sea Grant. 1988. *Fatty Fish Utilization: Upgrading from Feed to Food.* Raleigh, NC: UNC Sea Grant Publication 88-04.

von Brandt, A. 1984. *Fish Catching Methods of the World.* Farnham, Surrey, England: Fishing New Books Ltd.

Wheaton, F. W., and Lawson, T. B. 1985. *Processing Aquatic Food Products.* New York: Wiley.

Windsor, M., and S. Barlow. 1981. *Introduction to Fishery Byproducts.* Farnham, Surrey, England: Fishing New Books Ltd.

Sources of Information

Official Bodies

Food and Agricultural Organization of the United Nations, Rome: via delle Termedi Caracalla 00100. (North America: 1001 22nd St. N.W., Washington, D.C. 20437)

National Fisheries Institute, 1525 Wilson Blvd., Arlington, VA 22209

National Marine Fisheries Service, 1825 Connecticut Ave., N.W., Washington, D.C. 20235

Natural Resources Institute, Central Ave., Chatham Maritime, Kent ME4 4TB, England

Torry Research Station, PO Box 31, Aberdeen AB9 8DG, Scotland

Addresses of Publications and Publishers

Infofish, P.O. Box 10899, Kuala Lumpur, 01-02 Malaysia

National Fisherman, Seafood Business Report, 120 Tillson Ave., Suite 201, Box 908, Rockland, ME 04841-0908

Commerical Fisheries News, Box 37, Stonington, ME 04681

Meat Industry, 90 Throckmorton Ave., PO Box 1059, Mill Valley, CA 94942

Fishing News International, Seafood International, Fish Farming International, 81–89 Farrington Road, London EC1M 3LL, England

Quick Frozen Food, 7500 Old Oak Boulevard, Cleveland, OH 44130

Frozen Food Age, 230 Park Ave., New York, NY 10017

Pacific Fishing, 1515 N.W. 51st St., Seattle, WA 98107

Seafood Leader, 1115 N.W. 46th St., Seattle, WA 98107

Commercial Boating, 1995 N.E. 150th St., North Miami, FL 33181

The Sou'wester, PO Box 128, Yarmouth, NS, B5A 4B1, Canada

Canadian Fishing Report, PO Box 818 Station B, Ottawa, Ontario K1P 5P9, Canada

World Fishing, IPC Industrial Press Ltd., Quadrant House, The Quadrant, Sutton, Surrey, SM2 5AS, England

Fishing Gazette, 461 Eighth Avenue, New York, NY 10001

Lodestar, 805 W. Second Ave., Anchorage, AK 99501

Flashes, National Fisheries Institute, 1525 Wilson Blvd, Arlington, VA 22209

AFZ International, Chateau Amiral, Bloc B-42, Boulevard d'Italie, MC 98000, Monaco

The Fisherman, 623 Washington Avenue, PO Box 658, Grand Haven, MI 49417

The Inspection Connection, National Seafood Inspection Program, F/S32 NMFS, Washington, DC 20235

Water Farming Journal, 3400 Neyrey Dr., Metairie, LA 70002

Aquaculture Magazine, 31 College Place, Asheville, NC 28801

Food Processing, 301 E. Erie St., Chicago, IL 60611

Prepared Foods, 8750 West Bryn Mawr Ave., Chicago, IL 60631

Fishing News Books, Blackwell Scientific Publications Ltd., Osney Mead, Oxford OX2 0EL, England

Van Nostrand Reinhold (Avi and Osprey Books), 115 Fifth Avenue, New York, NY 10003

Index